A
Field
Guide
to the
Invisible

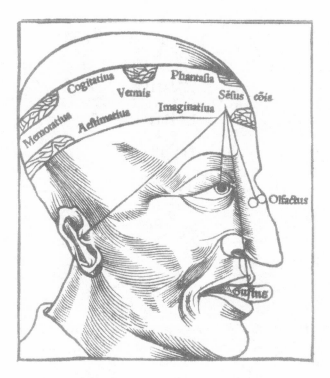

A
Field
Guide
to the
Invisible

Wayne Biddle

Henry Holt and Company
New York

Henry Holt and Company, Inc.
Publishers since 1866
115 West 18th Street
New York, New York 10011

Henry Holt® is a registered
trademark of Henry Holt and Company, Inc.

Published in Canada by Fitzhenry & Whiteside Ltd.,
195 Allstate Parkway, Markham, Ontario L3R 4T8

Library of Congress Cataloging-in-Publication Data

Biddle, Wayne.
A field guide to the invisible / Wayne Biddle.—1st ed.
p.   cm.
Includes bibliographical references and index.
ISBN 0-8050-5069-8 (alk. paper)
1. Science—Miscellanea.   I. Title.
Q173.B597   1998   97-39178
500—dc21

Henry Holt books are available for special promotions
and premiums.
For details contact: Director, Special Markets.

First Edition 1998

Designed by Betty Lew

Printed in the United States of America
All first editions are printed on acid-free paper. ∞

1   3   5   7   9   10   8   6   4   2

Consider, therefore, this further evidence of bodies whose existence you must acknowledge though they cannot be seen.

—Titus Lucretius Carus

· *Contents* ·

· *List of Illustrations* ·

· *Acknowledgments* ·

All field guides, even the ones with no tongue in cheek, are necessarily severe distillations. Almost every entry in this book represents lifetimes of study for devoted specialists. My pleasant job has been to marvel at what's been done, then to exit the dabbler's paradise and invoke a personal sense of importance. The bibliography is thus more than a formality. It is a bow to those who would spend years to produce a single volume entitled, for example, *Mites.*

I am indebted to Mimi Harrison for casting her professional eye on the choice of illustrations, in addition to showing considerable forbearance as the author's wife. Elizabeth Douglas of the Southwest Research Institute in San Antonio gave generously of her time to a supplicant on the telephone. I also warmly recognize the continuing, though sometimes financially tenuous, hospitality of research staff at the Library of Congress, the National Library of Medicine, the Milton S. Eisenhower Library at Johns Hopkins University, and the Gelman Library at George Washington University. To paraphrase Pierre Bosquet's observation about the Light Brigade's charge at Balaklava, the Internet is magnificent, but it is not scholarship.

Les cinq sens.

· *Introduction* ·

Much of everyday experience takes place beyond the range of our senses, which are ample but comparatively dull. It is our big brain that makes us superior to other creatures, for better or worse, and that also enables us to explore realms that are no less important because they are invisible.

Invisible means everything going around and through us that we not only can't see, but can't do much about anyway, at least as individuals. It suggests a mode for cataloging the inescapable stew we're in, for identifying life's ingredients that are literally out of sight and therefore too often out of mind. There is nothing here that you could easily ascertain on a hike through the woods, yet each item is as impossible to avoid as the trees. In our contemporary predicament, where so much seems beyond personal control, helplessness requires sorting out.

For example, an average adult breathes about 13.5 kilograms of air per day (but consumes only 1.2 kg of food and 2 kg of water). Since no one can go on an air diet, this figure represents a fixed universal. Also fixed, therefore, is exposure to more than two dozen gases—from the good, such as oxygen, to the bad, such as methane, to the ugly, such as formaldehyde. There is no clean air left anywhere on Earth, so we can no more avoid breathing the formaldehyde than the oxygen. What is invisible generates an index of what we are (in the Feuerbachian sense of "Der Mensch ist, was er isst"—man is what he eats), which is to varying degrees toxic whether you're a Democrat or a Republican, paranoid or not.

In medieval times, as William Manchester has noted, everyone knew perfectly well that the air was rife with menacing spirits—the souls of unbaptized babies, undead ghouls wandering forever between Heaven and Hell, sociopathic nymphs, melsh-dicks who made hazelnuts disappear, assorted incubi and succubi who did their dirty work while the victim's eyes were closed by sleep. If, moreover, any visible object might express invisible forces, then the senses were most useful for identifying telltale signs of the netherworld. Hence, perhaps, the priesthood as the highest intellectual calling.

Scientists supplanted priests long ago, of course, relegating holy or unholy ghosts to literature. Still, much of what is invisible to us sophisticated moderns is as scary as the enmity of old fiends. That our ancestors evolved to coexist with these realities is small comfort. So-called background radiation fits this category. If the intensity of ultraviolet rays reaching the Earth's surface had been harmful to life during the eons of human evolution, maybe we would have developed an ability to see them coming. But the planet's atmosphere, which largely filtered them out, mercifully made that unnecessary. Now that we have altered the atmosphere's chemical composition to the point where UV is a health threat, we must somehow grapple with an invisible foe. Or should only frogs be worried? Let's sort out such practical quandaries, which today are legion.

Julius Caesar is said to have banned nighttime chariot driving on Rome's cobbled streets because the noise spoiled his slumber. At the beginning of the fourteenth century, England's

Edward I forbade the burning of coal because he hated its smell, and executed at least one wretched violator. None of our leaders will ever possess the absolute power over the invisible enjoyed by these potentates. In a democracy, invisible threats are notoriously difficult to recognize and control because most of our lingua franca derives from visual evidence. The introductory phrase *let's see* is routinely applied to all the other senses: *Let's see if I can hear you whisper. Let's see if I can feel the tingle. Let's see if you have a fever. Let's see if we can taste the spice.* Seeing is believing for most voters, a positive trait unless it perpetuates states of denial. "This is hilarious," a frustrated citizen told the local paper about restrictions during a summer ozone alert around Washington, D.C. "It's just . . . it's just *oppressive.*" At least he wasn't beheaded for mowing his lawn, though we may someday come to that.

In this century, physicists have developed a special language, called quantum electrodynamics, for describing the unseeable interior of atoms. That it corresponds not one bit with everyday experience is irrelevant to them, given the power the concepts bestow. This is the age-old seductive pull of the invisible, which draws scientists down the rabbit hole just as ineluctably as it once drew priests. The professional language for comprehending invisible things like quarks, radiation, or ozone is often even more arcane than run-of-the-mill scientific jargon, widening the gap between those who see ("I *see*") and those who don't.

Everything is invisible for the blind, no matter what their education. So if blind people can still learn to step around holes, then the rest of us are evidently blessed with deep redundancies of data. The worst threats merely foil several layers, giving "odorless and tasteless" gases like carbon monoxide an especially demonic reputation.

In a more benign epoch, a field guide to the invisible might be construed as a spiritual exercise, and a demeaning one at that. "Why has not man a microscopic eye?" Alexander Pope mused in 1732. "For this plain reason: man is not a fly." But in an age of chronic, low-dose exposure to sundry radiations and pollutants; of drug-resistant infections from exotic microbes; of habitats where the sources of stress are amorphous; of a biosphere so radically changed by the hand of man that the natural protections it once provided are no longer assured, it is still the invisible that worries us most. Like superstitious sinners or bad scouts, we feel unprepared.

Then again, there are many good things among the invisible—friendly things whose invisibility enhances their marvelousness. As Emily Dickinson wrote: "So soft upon the Scene / The Act of evening fell / We felt how neighborly a Thing / Was the Invisible." In this niche we shall skip such unseeables as hummingbird wings or the rebounded sphere of a falling raindrop that are invisible because they pass too quickly. (Our nervous system can only distinguish between two moments if they are separated by more than about one-twentieth of a second.) A sane person does not see lightbulbs flashing on and off 360 times every minute. Being able to see all of creation would be a terrible curse even for the most crazed naturalist. Included here, nonetheless, are a few items that we might personally want to preserve as unseeable, though not unknowable. These deserve our surrender, actually, given how capable they are of giving pleasure. For the 40 percent of American scientists who call themselves believers, God herself may belong in this category.

What follows are points on the map of a parallel world, ignored at our peril, that we inhabit simultaneously with the one before our very eyes. There is a lot of difference between say-so and take-a-look, but we'll get over it.

A
Field
Guide
to the
Invisible

• allergens •

Hay fever (this colloquialism dates to the early nineteenth century, when POLLEN was first recognized as a culprit), or allergic rhinitis, is the single most common disease suffered by humans. The fact that we're not alone—apes get it, too—is no comfort. There are many irritating (and sometimes far worse) medical problems—skin rashes, itchy eyes, asthma—caused by hypersensitivity to otherwise harmless biological or chemical matter known in this context as allergens, but only hay fever has elicited enough literary bathos to inspire a play by Noel Coward.

Allergen was coined in 1912 after an Austrian pediatrician named Clemens von Pirquet formed the term *allergy* (or *Allergie*, in German) from Greek words meaning "abnormal response," no doubt referring to the over-the-top reaction of an allergic individual's immune system to certain invaders. Sicknesses recognizable today as allergic reactions have been reported for thousands of years. King Menes of Memphis is known to have expired in 3100 B.C. either from a hornet sting or from getting squashed by a hippopotamus (ancient Egyptian vocabulary being ambiguous on this subject). Intolerance of certain foods was discussed from the sixteenth century onward, as was the tendency

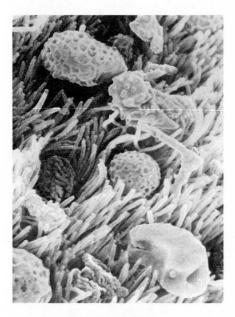

Allergens on the trachea's surface.

of pet dander, feathers, and other common substances to trigger asthma. In 1835, researchers began to "challenge" subjects with pollen samples.

All natural allergens are proteins or protein-carbohydrate compounds (glycoproteins in the jargon of organic chemistry). In the usual scenario, when any of myriad allergenic substances—animal dander, POLLEN, mold, FUNGI, BACTERIA, dust MITES, latex, cockroach mung, *et cetera ad infinitum*—land on us or in us, we all produce little soldiers that capture the foreigner and take him to jail. Speaking a bit more technically, proteins called immunoglobulin antibodies bind to the encountered antigens and then attach to receptors on mast cells, which are thereby stimulated to release inflammatory agents such as histamine (of snotty-nose fame, also contained in the hairs of stinging nettles) from their cytoplasm. Cat antigen, for example, is mostly from dried saliva. Unfortunately, lots of people are inclined by heredity to unleash way too much defensive stuff, sending nearby tissue into miserable symptoms of infamous category.

Why? No one knows. If there were some Darwinian advantage conferred upon itchy, rashy, sneezy, grumpy cavemen, the point is now lost on their descendants. There must have been a good reason because the genetic imperative is very strong: if one parent has allergies, children face at least a one-in-three chance of having them, too; if both parents do, the odds rise to seven-in-ten. More than 50 million Americans thus endure allergic diseases sometime during their lives.

Whoever classified ragweed as *Ambrosia* was obviously not hounded by allergens.

• alpha rays •

On a cloudy February day in Paris, 1896, Henri Becquerel put away an experiment he had been hoping to run that needed lots of sunshine. Forty-four years old and already sporting a long white beard, the École Polytechnique professor liked to tinker with uranium salts, which glowed wondrously in the dark after being out in the sun. His scientific coterie was *captivé* by Herr Röntgen's recent discovery of X RAYS at Würzburg's Physikalisches Institut in Bavaria, so Becquerel thought he might see if this stuff worked in the latest laboratory trick—placing a metal cutout between the uranium and a photographic plate to make a silhouette image. But it wouldn't work without sunlight to stimulate phosphorescence first, he believed. Everything went into a drawer to await clear skies.

Henri Becquerel, the impatient discoverer of les rayons uraniques.

Three overcast days later, Becquerel grew impatient and developed the plate anyway, maybe to prove to himself that the sun really was necessary. *Merde.* There as plain as his whiskers was a blotchy patch rather like a lipstick kiss. Something potent was coming out of the uranium salts even without the sun's prodding. After feverish study he christened the phenomenon *les rayons uranique*—uranium rays—and showed that they were different from X rays in two crucial respects: they could not pass easily through materials, and they streamed continuously out of the uranium. For the eternal glory of France, he had discovered radioactivity (which he still thought was a form of phosphorescence). For this feat he shared the Nobel Prize in 1903 with two other pioneers, Marie and Pierre Curie.

It was the inimitable Madame Curie who coined the term *radioactivity* and proposed that the rays were due to changes

Manya Sklodowska, the illustrious Madame Curie, in 1921.

within atoms, not to chemical reactions. At first ignored by Nobel nominators from the French Academy of Sciences, her work was championed by Pierre, who had put aside his own study of magnetism to collaborate with her. After he was killed by a horse-drawn wagon in 1906, she assumed his position on the Sorbonne faculty, raised two daughters (one of whom, Irène Joliot-Curie, shared the 1935 Nobel Prize in chemistry with *her* husband, Frédéric), and won another Nobel solo in 1911. Besides her scientific genius, she is remembered for a letter written to the Swedish Academy of Sciences, which was nervous about the very public scandal of her affair with a married physicist. "The prize has been awarded for the discovery of radium and polonium," she reminded them. "I believe that there is no connection between my scientific work and the facts of private life."

News of Becquerel's work had traveled quickly across the Channel to the Cavendish Laboratory at Cambridge, where a New Zealander named Ernest Rutherford was looking for a way to make his mark for the empire. In 1897 he began what would stretch into nearly ten years of radioactivity research. His first important find was that uranium rays had two components. One, which he called alpha radiation, was positively charged and completely absorbed by a sheet of paper or aluminum foil as thin as two-thousandths of a centimeter; the second, named beta radiation, was negative and 100 times more penetrating. We now know that all the natural elements heavier than lead, excepting bismuth, are radioactive and give off alpha or BETA RAYS.

The truly fabulous thing about alpha rays was that when one element emitted them, it turned spontaneously into a different element. Uranium, for example, transmuted into thorium. Shades of Merlin! "Don't call it *transmutation*," Rutherford warned a colleague, "they'll have our heads off as alchemists." Rutherford concentrated on alpha rays, concluding in 1908, the same year he won the Nobel Prize, that they were beams of particles like charged helium atoms. It took another two years to identify them as helium ions (clusters of two protons and two neutrons—not known at the time), actually *nuclei,* a new concept in atomic structure.

Along the way, Rutherford noticed that when alpha particles passed through an extremely thin sheet of mica onto a photographic plate, they did not make a sharp image; evidently they were somehow being scattered by the mica. But since they were moving at one-twentieth the speed of light (that is, with great energy), what was powerful enough to deflect them? More experiments with gold leaf just a few hundred atoms thick showed that sometimes an alpha particle even ricocheted backward off the target.

"It was as though you had fired a fifteen-inch shell at a piece of tissue paper and it had bounced back and hit you," he marveled. The only way this could happen was if each gold atom had a positively charged center that contained almost all of the atom's mass. If true, then atoms were more like miniature solar systems than the "plum pudding" model then in vogue. He calculated that the center, or nucleus, occupied less than 10^{-14} of the atom's volume. (If Earth were an atom, its nucleus would be the size of Yankee Stadium.) Matter was thus mostly emptiness. The Danish physicist Niels Bohr, after working in Rutherford's lab, developed the theoretical aspects of this planetary model, which came to be named after Bohr.

Rutherford—now Sir Ernest—announced his last great discovery in 1919 at the age of forty-eight, after the Great War had dispersed many of his international colleagues. He found that the positive charge in the nucleus was due to particles he called protons, from the Greek word for "first." Not a bad career for a boy from Nelson. He was named director of the Cavendish, despite his colonial status. His bread-and-butter technique of firing alpha rays at various targets and analyzing the reflections paved the way for "atom smashers," which are still the most powerful experimental tools of physics.

Alpha rays in nature are generally harmless to people because the first layer of dead cells in our outer skin is too thick for them to penetrate. Nor can they make it to the lens of the eye. Natural and man-made emitters include radium, plutonium, thorium, neptunium, americium, and curium. Very few alpha particles can travel more than one to three inches in air before being stopped. Polonium-212 and 214 give off alpha rays that are energetic enough to reach inner cells, but 212 is very rare and 214 is of concern only in uranium mines, where it is a suspected cause of skin cancer among miners. Therapeutic beams of alpha particles have been used instead of X rays to treat eye cancers and certain brain disorders because the cell destruction they cause is more confined.

If the elements that emit alpha rays somehow get inside us, however, they can be very damaging and potent carcinogens as the particles travel through soft tissue and destroy chemical bonds in cellular molecules (see RADON). A dose of alpha radiation delivered to living tissue is twenty times more destructive than the same dose of X rays. Radium-226 (the isotope studied by Marie Curie) was one of the first substances tied conclusively to cancer, in classic epidemiologic studies of watch-dial painters. In 1931, researchers reported that young women employed to paint luminous numbers on watches and clocks developed bone cancers far more often than the general population because they often "tipped" their brushes with their lips. Due to its chemical similarity to calcium, the swallowed radium accumulated in their bones, where emitted alpha rays produced tumors. (Today, tritium—which gives off only low-energy beta rays—has replaced radium in timepieces.)

During and after World War II, military doctors treated inflamed inner-ear tissues of submariners and airmen with direct exposure to radium. The technique, called nasopharyngeal radium therapy, involved pushing a rod filled with fifty milligrams of radium through the nostrils until it reached the eustachian tubes, which help balance pressure. Some 20,000 troops and dependents endured the procedure for inner-ear problems from the 1940s to the mid-1960s. A 1982 study by Johns Hopkins researchers found that they might be at higher risk for head and neck cancers, but the Pentagon did not publicly report the matter until 1997.

In 1934, Madame Curie herself died at age sixty-six of leukemia induced by decades of radiation exposure.

• **ammonia** •

A colorless, poisonous gas that fortunately has a pungent odor irritating enough to make us get away from it fast. Think of smelling salts—a dilute mixture of ammonia and PERFUME still powerful enough to bring a KO'd heavyweight back to his senses. The large-scale synthesis of ammonia for nitrogen fertilizer is the keystone of modern agriculture and thus of global coastal water pollution.

The standard manufacturing method of combining atmospheric hydrogen and nitrogen under high pressure to make NH_3 is still called the Haber-Bosch process, after the German chemists Fritz Haber and Carl Bosch, who pioneered it just before World War I. The illustrious Haber, founding director of the Kaiser Wilhelm Institute for Physical Chemistry in 1911, received the Nobel chemistry prize in 1919 for this synthesis, despite the fact that his work had allowed Germany to keep mass-producing explosives after the Allies blocked shipments of nitrate from Chile. The Swedish Academy, focusing on man-made ammonia as a boon to world agriculture, also chose to ignore his pivotal role in developing deadly battlefield gases for the *Wehrmacht,* for which the Allies considered him a war criminal. (Just before the first such use of gas at Ypres in

1915, his wife committed suicide.) Bosch went on to receive a Nobel in 1931 for further research on high-pressure chemistry. Haber, who had always been a fervent patriot, left Germany when the Nazis came to power in 1933, because he was Jewish. He died the following year in Switzerland.

Vast quantities of ammonia are used today for making fertilizers, nitric acid, plastics, explosives, dyes, glues, pharmaceuticals, petroleum products, synthetic fibers, and myriad other chemicals. Total abundance in the biosphere is esti-

Fritz Haber, Nobel chemist and war criminal.

mated to be 10 billion tons. At concentrations over 20 to 25 parts per million in air, surveys of workers have found symptoms starting with nasal dryness (32 ppm) and eye, nose, throat, or chest irritation (72 ppm). Industrial accidents with anhydrous (water-free) ammonia at inhalation exposures of 2,500 to 6,500 ppm have caused blindness and fatal lung damage. The federal limit is 25 ppm.

About 45 percent of the NH_3 in the atmosphere comes from decomposition of animal waste, including human excreta. In Europe and Great Britain, cow piss is the number-one source via urea, $CO(NH_2)_2$. Urban and industrial wastewater treatment often generates ammonia pollution. The action of BACTERIA on organic matter in soil accounts for another 20 percent of atmospheric ammonia, some of which is reabsorbed by plants as they try to keep the concentration of NH_3 around them constant. Release from fertilizers produces about 8 percent, and burning of coal less than 1 percent. Biomass burning can be significant in the tropics, but most of the rest comes from the oceans. On a worldwide basis, about 54 percent of atmospheric ammonia comes from human activities; over continents the fraction reaches almost three-quarters.

The world's reliance on nitrogen fertilizers, underpinned by factory production of ammonia, appears to be absolute. At least 2 billion of the planet's 6 billion people are able to exist only because of plant and animal proteins derived from nitrogen supplied through Haber-Bosch synthesis. Application of these fertilizers in the United States has stabilized since the 1980s, but ecologists estimate that half of all industrial nitrogen ever used worldwide has been spread on the ground since 1984. Since 1993, a "dead zone" of hypoxic (oxygen content below 2 ppm) water has appeared every summer across some 7,000 square miles of the Gulf of Mexico along Louisiana's coast, as algae bloom on hundreds of millions of pounds of nitrates in debouched Mississippi River sediment. A similar "eutrophocation" (nutrient enrichment of oxygen-depleted water) disaster plagues Chesapeake Bay, where *Phiesteria piscicida* microbes that thrive in nitrate-rich stagnant waters caused massive fish kills and human illness along the Pocomoke River in 1997; Scandinavian fjords; and coastal waters on every continent except Antarctica. Since 1972, the number of major shoreline outbreaks of algal diatoms and plankton harmful to fish and higher forms of life—including humans—has more than doubled in the United States, from sixteen to thirty-six. The Environmental Protection Agency has set no federal pollution standards for nutrients such as nitrogen beyond the drinking water limit of 10 ppm, which is far above what will foster hypoxia.

Ammonia is used by cigarette manufacturers to boost the biological effects of nicotine. In natural tobacco, nicotine resides in molecules of nonvolatile salts. Ammonia releases pure nicotine from these compounds, which evaporates more easily and dissolves better in fats. When smokers inhale, the addictive drug is then more readily absorbed by lung tissue. Cigarette tobacco typically contains about 3 percent ammonia, which helps evaporate about a quarter of the nicotine in inhaled smoke—about 100 times more than without ammonia. According to a 1991 tobacco company "leaf-blender's manual" obtained by the Food and Drug Administration, this is "associated with increases in impact and 'satisfaction' reported by smokers." The process is similar to "freebasing" cocaine, where the snow (also a salt, cocaine hydrochloride) is liberated by ammonia or, to make crack, by sodium bicarbonate.

• **background radiation** •

Nature is radioactive, including us. Edward Teller, father of the H-bomb, loved to emphasize that we are all intimately irradiated by the potassium-40 in our bedmate's body, that by snuggling with two lovers we would get more radiation than from lounging against a nuclear reactor.

Just because we sleep with background radiation doesn't mean it's harmless, Dr. Strangelove. Even today, some textbooks repeat the shibboleth that the highest doses to people living near the Three Mile Island accident in 1979 were less than the annual background exposure and thus nothing to worry about. Like germs (which it somewhat helps keep at bay), background radiation causes health problems—just not often enough to be highly noticeable in a crowd.

People living with sharply above-average natural radiation in parts of Brazil, China, and India, due to thorium-rich monazite soil (beaches in southwestern India are streaked with sand containing as much as one-third thorium), have more chromosome aberrations and, in the last case, greater incidence of Down's syndrome than control groups. Higher frequency of cancer has not been documented, but the epidemiologic studies suffer statistically from low numbers of subjects. Villagers near Rio de Janeiro received radiation detectors in the shape of religious medals to bolster their participation.

Where does natural background radiation come from? There are two general sources: cosmic and terrestrial. COSMIC RAYS account for about 9 percent of the total annual exposure. A passenger on a cross-country airline flight receives about 2.5 millirems from this source. RADON is by far the largest earthbound contributor, at about 66 percent. So-called primordial nuclides—potassium-40, rubidium-87, and members of the decay chains of thorium-232 and uranium-238 (plus, to be obsessive, the extremely rare uranium-235)—left in the Earth's crust when it cooled from radioactive gases supply another 9 percent. The rest comes from deposits inside everyone's body, mainly potassium-40 and rubidium-87. The dose of GAMMA RAYS from your spouse's potassium-40, which is the only type of internal radiation capable of leaking out, is negligible compared with the emission of BETA RAYS from your own nuclides. So keep on spooning.

And maybe don't pig out on Brazil nuts, which are grown in high–gamma ray soil and may be 14,000 times more radioactive than most fruits. Other "hot" foods include cereal, organ meats, peanuts, and flour. Oysters have been known to accumulate zenon-65 to levels hazardous for gourmands. Eating tuna and wild salmon may increase the body burden of iron-55. High levels of strontium-90 have been found in the shells, bones, and edible parts of mollusks. As with contamination from pesticides and other food contaminants, government standards are only as good as enforcement, which—*we're shocked*—is sometimes spotty.

After the discovery of X RAYS in 1895, man-made radiation was added to Mother Nature's. Today this accounts for about 18 percent of the total yearly exposure from both natural and artificial sources, breaking down to 11 percent from diagnostic X rays, 4 percent from nuclear medicine, and 3 percent from consumer products. The last category is dominated again by radon (in domestic water lines) but also includes phosphate fertilizers, building materials such as cinder blocks, color video monitors, and, most notably, cigarettes, which will deliver a nasty little dose of ALPHA RAYS to your bronchial epithelium from polonium-210. Fortunately or unfortunately, depending on your personal thrill quotient, the fabulous shoe-fitting X-ray fluoroscopes of the 1950s are no longer the neatest thing about getting new shoes. Baby boomers who survived those sales gimmicks may recall the 1959 scare involving contamination of calcium in milk by strontium-90 from A-bomb blasts. Today, the contribution of such FALLOUT is relatively insignificant, assuming no more atmospheric tests.

All told, the average annual personal dose to the U.S. population from GOD and Mammon is about 360 mrem (or 3.6 mSv in the newer "Sievert" units, where 1 Sv = 100 rem). By being prudent about medical X rays and diligent about household radon, plus not smoking, you can reduce this figure somewhat as an individual. The rest, for better or worse, is just part of being an Earthling.

At the level of charlatanism, yet relentlessly pushed by the nuclear industry, is the subject of *hormesis*—the supposed healthy effect of low-level, whole-body radiation. Alpine "radon spas" have been selling this pseudoscience to true believers for 600 years. Some early animal experiments claimed that such exposure brought longer life expectancies, but the data were flawed. Human mortality from cancer and heart disease is lower in parts of the United States having high natural background radiation, but cause-and-effect is confounded—as in many "ecological correlations"—by other factors, such as the fact that these regions are mostly at high altitudes.

Let us end with more pearls of wisdom from Dr. Teller, who had the chutzpah to write a book in 1962 titled *The Legacy of Hiroshima:* "A person living in a brick house . . . is exposing himself to . . . perhaps as much as ten times the amount of the current dose from radioactive fallout. If fallout really is dangerous, we should tear down all of our brick houses. I would hate to do this, because I live in a brick house myself."

• bacteria •

Earth's dominant form of life. We have evolved to accommodate *them*. Back up 3.5 billion years, and the fossil record begins with bacteria. Their cumulative mass since then would be greater than the whole planet's, which is a mere 6.588×10^{21} (6 sextillion, 588 quintillion) tons. Their population in a single human gut far exceeds the total number of people who have ever lived. Louis Pasteur was so unnerved by their ubiquity that when dining out he used a magnifying glass to scrutinize his meal.

Scientists have identified more than 4,000 species of bacteria, with untold millions more to go. Most are harmless to us, many are helpful. About 10 percent of our body weight consists of them, some of which we absolutely depend on. The pseudomonads that swim on any recently moist surface would be quite sufficient to decompose every errant flake of organic matter in the house. Despite pasteurization, millions live in a pint of fresh milk. The invigorating smell of rich garden soil comes from gases released by streptomycetes, whose chemical defenses against other microbes we use in the antibiotics streptomycin and tetracycline. If we disappeared as a species tomorrow, it would make less than a blip in bacterial ecology. If just a trivial clan of their kind departed, like *Escherichia coli,* we'd be goners, too.

Bacteria are single cells, usually about one to two microns (thousandths of a millimeter) across. (The flecks of DUST visible in a sunbeam measure down to about twenty microns.) But some are as small as 200 nanometers (billionths of a meter). They come in three main shapes, called cocci (balls), bacilli (rods), and vibrios (commas), though you can find corkscrews, Ss, and squares. They reproduce asexually by just dividing in two.

Those of us who grew up believing that the 281,000 animals, 284,000 plants, and 750,000 insects described by taxonomists constitute the peak of biodiversity were demonstrating what happens when "the error of our eye directs our mind," as Shakespeare put it. The genetic richness of microbes was invisible to science until gene-typing techniques allowed microbiologists to see that groups of bacteria are as distinct from each other as dogs are from rosebushes. Multicellular animals did not appear until about 580 million years ago on the tree of life, which thus owes most of its limbs to bacteria.

They inhabit every conceivable nook and cranny suitable for life—generally wherever there is water from about 20° to 250° F, but not necessarily. Aerobes need air, anaerobes don't. Thermophiles want heat, psychrophiles like cold. Acidophiles need acidic milieus, alkalophiles desire alkaline. Heterotrophs eat organic matter, autotrophs crave CARBON DIOXIDE. They love glaciers, hot springs, and volcanic fissures, besides the usual scummy places we're used to finding them in. Bacterial SPORES have been collected around deep-sea sulfide "smokers" at 650° F, while others can remain viable for years despite desiccation. They have been extracted from a Swedish bore hole almost four miles deep. Lithoautotrophic bacteria thrive in rocks thousands of feet below the Earth's surface, using energy from inside the planet instead of the solar power of ecosystems we've always deemed conventional. In fact, microbes may have been the primary architects of the Earth's crust as they metabolized metals, minerals, and hundreds of other compounds. If the estimated 100 trillion tons of bacteria living underground right now were mined out and slathered across the Earth's entire surface, we'd be up to our necks in them.

Plants may be responsible for maintaining oxygen in the atmosphere, but this life-giving gas was supplied primordially by bacterial photosynthesis. Plants themselves could not survive without nitrogen-fixing bacteria living free in the soil (cyanobacteria) or symbiotically within root nodules (of the genus *Rhizobium*). These bacteria split apart tightly bonded nitrogen molecules and create AMMONIA, a compound whose nitrogen is easier for plants to incorporate. Methanogenic bacteria in the guts of cattle produce almost one-third of the atmosphere's METHANE, a crucial greenhouse gas that controls global temperatures. In fact, carbon, oxygen, nitrogen, sulfur, phosphorus, and every other element pivotal to life are churned throughout the biosphere by bacterial action.

Now for the downside. Of the millions of different bugs out there, perhaps a hundred are known to threaten our health when their parasitic colonies grow too populous (*E. coli* in the intestines, for example), or when they release powerful toxins (*Clostridium tetani* in a wound). The range of human reaction to these microbes is as broad as the number of people exposed to them, hence the physician's art. The thrust of evolution is toward peaceful relationships. Germs have no self-interest in wiping us out; they just want a well-adapted place to live. Some

600 kinds inhabit our mouths, for example, in what would be a mutually happy home if not for dietary excesses that jump-start *Streptococcus mutans* colonies, which love to turn sugar into acid. *Bordetella pertussis,* the whooping cough bug, can be found in perfectly healthy throats, as can *Mycobacterium tuberculosis.* Or they can kill us.

Our natural immune system polices all bacterial encounters, a.k.a. infections, but when it is overtaxed we get sick. To help fend off the very few kinds of bacteria with which we still have a rather savage repartee, medical technology has refined antibiotic drugs from their natural enemies, such as the *Penicillium* mold that Alexander Fleming accidentally discovered could kill staphylococcus. Evolution is a powerful express train, however, when trillions of cells are reproducing in generations that last only minutes or hours. The germs have thus become resistant to many frontline medicines prescribed profligately for the past fifty years.

In an affluent yet germophobic society such as ours, the most important thing to understand about bacteria is that they cannot be escaped, household disinfectants notwithstanding. A normal immune system will handle most of the pathogenic bugs, albeit with some chance of side effects otherwise known as symptoms of disease (which are treatable more often than not). Getting sick should be devoid of social meaning, since nothing is more natural than meeting a microbe. It's their scene. We must try to get along.

• bad breath •

"Not Homer's Chimaera breathed such foul breath, not the fire-breathing herd of bulls of which they tell, not all Lemnos [whose women were cursed with a stench by Venus, goddess of fragrance] nor the excrements of the Harpies, nor Philoctetes' putrefying foot." Thus did the epigrammatist Lucilius describe the breath of Telesilla. Madison Avenue would not foist the term *halitosis* on society for another 2,000 years, but bad breath was already a grave offense.

The high rank of bad breath among twentieth-century social ills traces directly to a businessman named Gerard Lambert, whose father developed Listerine in the 1870s as a general-purpose antiseptic. A rich boy who was whisked from class to class at Princeton in a chauffeured limousine, Lambert determined after graduation in the early 1920s that the family enterprise had grown stale profit-wise. A minion saw the word *halitosis* in a British journal, and Lambert seized upon it as the magic word that would cut a new market niche for Listerine. Saturation magazine advertising blamed unpleasant mouth odor for failure in romance ("The secret her mirror held back concerned a thing she least suspected—a thing people simply will not tell you to your face") and in business ("Fired—and for a reason he never suspected"). As a result of the ad campaign, annual profits rose from $115,000 in 1921 to $8 million in 1928, whereupon lucky Lambert sold his stock for $25 million just before the Crash. In his wake he left a legendary American success story, a touchstone for future hucksters, and a term that still stands for fearful obsession with our natural SMELLS.

Scientifically speaking, the oral cavity is a well-established source of both endogenous (from the lungs) and exogenous (from oral BACTERIA) odors. Concentrations of rotten-egg-smelling HYDROGEN SULFIDE may reach 700 parts per billion and of skunky methyl mercaptan nearly 200 ppb. Tooth cavities, periodontal disease, and tonsillitis generate a fetid smell and bad taste. Colds and sinus infections also bring on malodor.

Liver failure causes "fetor hepaticus," a mousy breath from methanethiol and other alkylthiols. Kidney disease makes breath smell like piss, and a lung abscess turns it putrid. Gastrointestinal disorders do not cause bad breath, so breath odor offers no hint of what's happening way down in the stomach or bowels.

In the Middle Ages, those few members of the gentry in a position both to care about their breath and actually do something about it chewed herbs or gargled with spiced water. Candies called "kissing comfits" might palliate one of the noxious olfactory hazards of intimacy, though in an era when bathing was considered a health threat, gross breath would have been a relatively minor consideration.

• beta rays •

In 1897, Ernest Rutherford detected and named two constituents—alpha and beta—of the mysterious "uranium rays" discovered a year earlier by Henri Becquerel (see ALPHA RAYS). Rutherford then turned his prodigious attention to alpha rays, while other physicists concentrated on the betas. Among them were Marie and Pierre Curie, who in 1900 found the electrical charge of beta rays to be negative; Becquerel himself, who determined that beta rays were similar to so-called cathode rays (which nowadays produce the image in television picture tubes), already known to be composed of electrons; and Walter Kaufmann of the University of Göttingen, who in 1902 proved that beta particles were electrons.

Physicists had known since 1897 that atoms contain electrons, so they were not exactly shocked by this news. (Not yet, that is. It would take quite a few more years to understand how beta electrons spew from a radioactive atom's nucleus, which doesn't contain electrons any more than a cow contains a *moo*.) In 1907 at the University of Berlin, Lise Meitner and Otto Hahn—who were fated to introduce nuclear fission just before World War II—decided to try measuring the energy of the emitted electrons, figuring that they would all be the same from any given radioactive element, as was true for alpha particles.

After some initial experiments, they published an article to this effect in 1908, but soon feared that their conclusions were wrong. By 1911, a new experimental technique suggested to them that beta electrons came out with a range of energies instead of just one. Other researchers—notably Rutherford's ace student, James Chadwick—took up the question, leading to a genuine crisis. It turned out that if one radioactive atom transmuted ("decayed" in the vernacular) into another while giving off a beta electron, the total energy before the change was not equal to the total energy afterward. In other words, the sacred principle of energy conservation appeared to be violated. Physicists were so confounded by this problem that even the great Niels Bohr was tempted just to accept it as a bizarre fact of life within atomic confines. Thus the stage was set by the late 1920s for the postulation of a bizarre new particle to account for the missing energy, the NEUTRINO.

Because beta particles have only 1/7,500 the mass of alpha particles and travel close to the speed of light, they are more penetrating than alpha rays. A sheet of newspaper will stymie an alpha particle, but a beta electron can pass through three feet of concrete. Compared with alpha radiation, beta rays are somewhat less damaging to biological tissues, since they tend more often to miss cellular molecules entirely. One of the most important beta emitters for human exposure is the naturally occurring radioactive isotope of potassium, technically noted ^{40}K, which is ubiquitous in plants and animals. For a given dose of radiation absorbed in living tissue, beta particles cause about the same injury as X RAYS or GAMMA RAYS.

Unlike alpha rays, beta particles from many radioactive substances can penetrate the epidermis. If concentrated beta emitters—such as the fission products in atomic bomb FALL-OUT—come into contact with skin, they can cause a disfiguring form of radiation injury called "beta burn." Marshall Islanders exposed to fallout from the American H-bomb test at Bikini atoll in 1954 suffered from extensive beta burns, especially because they wore little clothing that might have attenuated the radiation. The crew of a Japanese fishing boat named *The Lucky Dragon* was also seriously burned. In this ghastly case, fallout took the form of white lime powder from the decomposition of coral by blast heat. Some victims later developed thyroid cancer. Welcome to the atomic age, Bloody Mary.

· B.O. ·

"Why has the armpit a more unpleasant odor than any other part of the body?" asks Aristotle in his *Problemata*. We still don't really know.

On the other hand, consider Robert Herrick's "Upon Julia's Sweat": "Wo'd ye oyle of Blossoms get? / Take it from my Julia's Sweat: / Oyl of Lillies, and of Spike, / From her moysture take the like: / Let her breathe, or let her blow, / All rich spices thence will flow." Henri III of France sniffed the linens of Mary of Cleves after she changed them in a closet, thereby falling forever in love. Goethe went so far as to steal the bodice of Madame von Stein so that he could smell it whenever he had a few idle, Stein-less moments.

Obviously we're on subjective ground here. The human body excretes or secretes thousands of volatile chemicals, many with strong odors. We leave behind a cloud of bitter AMMONIA, rotty HYDROGEN SULFIDE, vinegary acetic acid, pungent ethyl alcohol, and skunky mercaptans. We are short work for bloodhounds. The stuff exudes from the most salacious natural gunk, including sweat, urine, feces, saliva, come, and skin oils. A potent man's aura seminalis was once proof of his virility. The mix reflects daily environmental influences such as drinking water, diet, drugs, and hygiene products, as well as personal factors such as gender, age, race, reproductive state, health, and emotional condition. "B.O." can be ghastly or sublime, even both at once.

If we limit our fascination here to skin secretions, we have to consider three possible culprits. First, there are the eccrine glands, which are distributed over almost the entire body and produce a sweaty solution of inorganic salts (potassium, sodium, zinc), glucose, vitamin C, riboflavin, and amino acids primarily to help regulate the body's temperature through evaporation. About two liters of liquid an hour can be produced by some 2 million of these sweat glands. They're generally considered nonodorous, but some foods—notably garlic—can change that in a hurry, as can certain metabolic disorders such as diabetes.

Second, the sebaceous glands secrete most of the lipids that keep the skin supple and waterproof. Their greatest density is on the forehead, face, and scalp (hence all that pancake makeup on talking heads), with none at all on the palms or soles of the feet. Specialized sebaceous glands are stationed at all points of entry into the body, including the eyelids, ear canals, nostrils, lips, nipples, and anogenital orifices. This is quite handy, since they also produce substances that retard the growth of pathogenic microbes. Though they're associated with acne, their secretions are thought to have a weak and not unpleasant odor.

Finally, the apocrine glands let loose a stew of compounds, especially during times of excitement or stress. They are thus the major source of B.O. in humans, found primarily in the armpits, but also in the sternal region of the chest, the anogenital area, the breast areola, the scalp, cheeks, eyelids, and ear canals. In the pits, bacteria thrive on the secretions to produce acrid-smelling isovaleric acid. The apocrines operate only between puberty and menopause and are much larger in males than in females (though women may have more of them), leading scientists to wonder if they have pheromonal functions. They also differ in size and number between various races (blacks have more than whites, who have more than Asians), leading to all sorts of ugly stereotypes.

Two of the most interesting apocrinal secretions are steroids (androstenone and androstenol) that also happen to be found in pigs, where they help produce the "tainted breath" of boars that stimulates sows to go into mating position. Since the 1970s, numerous behavioral studies have tried to establish whether underarm odors communicate sexual identity and are therefore preferred by males or females. Valiant test subjects have compared dirty T-shirts, sniffed gauze pads taped inside armpits, rated photographs while wearing surgical-type masks soaked or not soaked with androstenol, and, in one truly inspired experiment, they chose among chairs in a dentist's waiting room that had been sprayed with various concentrations of androstenone (women went for the smelliest seats, oh *yes*). Methodological shortcomings have usually undermined this search for a holy pheromonal grail.

In our deodorized civilization, the answer to B.O. is chemical counterattack. Sweat beads are negatively charged under the positive skin pores, so that by smearing negatively charged

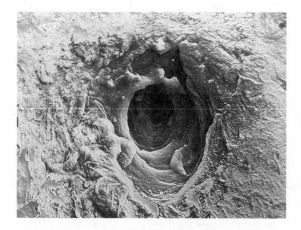

Mmm . . . sweat pore.

bits of aluminum across an armpit, we neutralize the magnetic force that draws perspiration to the surface. Add a little PER-FUME and some pesticide to the concoction, and you've got a product worth billions.

• **burps** •

Swallowed air, referred to in uptight circles as *aerophagia* or *eructation*. Everyone gulps a bit of atmosphere as they eat or drink, then belches it back up either privately or with gusto, depending on the social milieu. Anxiety sometimes creates additional, unconscious intakes. So may chewing gum or smoking. As Mother always suspected, table manners help decide which way the air ultimately goes, perhaps more than what's being consumed (soda pop and beer being well-known gas bombs). A Teutonic posture keeps it above the stomach contents, where it rises without much trouble besides the auditory insult. Hunching over like a dog tends to trap it below, where it may get forced into the duodenum and contribute to FLATUS. Sit up straight! Listen to your elders!

• **carbon dioxide** •

CO_2 is a harmless trace gas in the Earth's natural atmosphere, the fifth most abundant substance in unpolluted air after nitrogen, oxygen, water, and argon. It's the fizz in fizzy water and in solid form makes "dry ice." In the presence of moisture it forms carbonic acid, which has pitted the visage of marble statues from Athens to Anaheim. It is vital to the process known as photosynthesis—and thus to all life—in which green plants, algae, and some BACTERIA use the energy of sunlight to make nutrients. It also helps regulate the planet's temperature, which is why it may soon cause *big-time* problems.

Aristotle took the first conceptual leap toward understanding photosynthesis. (He was thinking about how plants grew and did not use this term, of course, which was coined in 1898.) But not until the late eighteenth century did science begin to provide real insights. In 1771, the English theologian and pioneer chemist Joseph Priestley (1733–1804) found that air inside a jar changed somehow if a rat was in there breathing it for a while. He called the different gas "phlogistic matter," which we now know to be exhaled carbon dioxide. He must have discovered simultaneously how to make rats faint, but he is not famous for that.

More to the point, the jar's contents could be "dephlogisticated" by replacing the rat with a plant. Priestley's direct link to the ancients is revealed by his jargon—*phlogiston* was the supposed substance of fire. A candle flame, he observed, was immediately overwhelmed by the mysterious "phlogistic matter." Science it was, albeit only a step or two out of the alchemist's den. Perhaps this is why, besides his challenge to established religious order as founder of English Unitarianism, Priestley's church was ransacked by a mob and his books burned in Dort, the Netherlands, by the town hangman.

Photosynthesis, summed up by the notation $CO_2 + H_2O \rightarrow O_2 +$ [organic compounds], is one of the great global engines that make life as we know it possible. On a worldwide scale, the production of O_2 is almost perfectly balanced with its consumption via respiration. (Fortunately, O_2 has such a long residence time in the atmosphere that even if Lex Luthor managed to halt all photosynthesis, it would take more than 8,000 years to use up all the existing oxygen.) Priestley's personal faith was no doubt deepened by glimpses of a cycle so seemingly infinite and self-sustaining. What could have set it in motion besides the hand of GOD? Our unfortunate lot today is to appreciate that nature can be undone by the hand of man, hence the second important thing to comprehend about carbon dioxide.

During the International Geophysical Year of 1957—the same *Wissenschaftfest* that brought us Sputnik—researchers built a CO_2 measurement station on the 9,000-foot peak of Hawaii's conveniently dormant Mauna Loa volcano. Sixty-one years had passed since a Swedish scientist named Svante Arrhenius first calculated how changing the concentration of carbon dioxide in the atmosphere would affect the temperature of the Earth's surface. When solar energy reaches the planet, some is reflected by clouds and about 7 percent is absorbed as ULTRA-VIOLET radiation by OZONE in the stratosphere, but most reaches the surface as visible light and then radiates back out into space as INFRARED energy. About 30 percent of this heat is trapped by carbon dioxide and WATER VAPOR in the troposphere, which keeps the surface far warmer (by about 54° F) than otherwise. Hence the veddy English name "greenhouse effect," which dates to research carried out in the United Kingdom during the 1930s.

In 1939, an article titled "The Artificial Production of Carbon Dioxide and Its Influence on Temperature" was published in *Quarterly Journal of the Royal Meteorological Society* by George Callendar, who realized that burning massive quantities of carbon-based fuels like coal was changing the natural balance of gases in the atmosphere. Twenty years later, the Mauna Loa station provided data showing that CO_2 was steadily rising by 0.2 to 0.3 percent per year—from about 315 parts per million in 1958 to more than 350 ppm in the 1990s. Other monitoring points around the world have confirmed this trend.

From analysis of air bubbles trapped inside deep ice core samples, we now know that the preindustrial (before 1750) concentration was around 275 to 285 ppm, with the steepest rise happening in this century. Somewhere between 90 and 145 billion tons of carbon have been added to the atmosphere since 1850. "Within a few centuries, we are returning to the atmosphere and oceans the concentrated organic carbon stored in sedimentary rocks over hundreds of millions of years," two leading researchers warned even before getting the bad news from Mauna Loa.

The biggest culprit is the internal combustion engine, followed by power stations, cement kilns, and aircraft. Tropical deforestation has exacerbated the problem. Auto manufacturers continue to campaign hard against mandatory CO_2 emissions controls, which require improving fuel economy or simply cutting the number of cars on the road. "Almost 97 percent of CO_2, after all, comes from natural sources such as plants and oceans," argues Chrysler chairman R. J. Eaton, waving one of the debate's great red herrings. The Big Three propose a joint development project with national laboratories to create a "Supercar" that would get eighty miles per gallon with less than half the CO_2 exhaust of current vehicles. At a time when gas-guzzling "sport utility vehicles" are hot sellers and alternative motors that run on natural gas, alcohol, or electricity remain expensive novelties, the political outlook for air-quality regulations is as scrambled as Eaton's logic.

Expressed in terms of a tax on fossil fuels, the Clinton administration estimates the cost of keeping emissions at 1990 levels by the year 2010 would be like raising the tax on gasoline by about twenty-six cents a gallon, on electricity by two cents per kilowatt hour, on coal by $52 per ton, and on natural gas by $1.49 per thousand cubic feet. Of course, other economic methods to reduce greenhouse gas emissions—subsidies for energy efficiency, for example, or a trading system in credits for cutting pollution—would be more palatable than taxes.

Of obvious concern to American industrialists is the fact that a United Nations global-warming treaty to maintain 1990 fossil fuel usage would exempt 130 developing nations, including China and India. Jobs would follow cheaper (that is, dirtier) energy. So far this decade, carbon dioxide emissions have increased 7.8 percent in OECD countries. In the United States, industrial CO_2 pollution now exceeds pre-1980s levels. With just 4 percent of the world's population, the United States belches out more than a quarter of all greenhouse gases. The fifty largest electric companies in the eastern half of the United States account for 20 percent of the nation's carbon dioxide emissions, led in 1995 by Kentucky Utilities.

In the decades between 2020 and 2080, depending on various postulations about ocean-atmosphere dynamics and how much the world continues to rely on carbon-based fuels, the CO_2 concentration could double from current values. This would produce global warming of 1.8° to 6.3° F by 2100, amplified by a factor of three or four in polar regions. The effects of burgeoning carbon dioxide are not yet plainly noticeable, given how much the weather fluctuates normally from year to year. Researchers agree that the average ground-level temperature increased by about 1° F over the past century, with a more pronounced rise of 1° to 3° F in the northeastern United States. But most global climate models, based on reasonable assumptions, predict that average surface temperatures will increase far more over the next 100 years. (The models cannot account for the fact that middle portions of the troposphere have actually been *cooling*—a glitch that makes greenhouse skeptics smile.)

Under such conditions, leaving shoreline property to your grandchildren might be very bad estate planning. Cape Cod would lose some 200 acres a year as the sea level rises twelve to twenty inches. The top of the Pilgrim Monument in Provincetown could house the last bar in town. New England would suffer more droughts, possibly leading to a prairie environment like that of eastern Kansas. Earlier springs and hotter summers would alter which tree species prosper—stifling the beloved sugar maple, for example—and cause mosquito population booms. Best long-term investment? Canadian wheatlands, which will replace parched American farms as the world's breadbasket.

The demand for ice-cold carbonated beverages would skyrocket, too, of course, perhaps suggesting a use for all that excess gas if it could be siphoned from the atmosphere. CO_2 stimulates the trigeminal nerves of the tongue, causing a prickly sensation that borders on pain at high enough concentrations. In the 1770s, Joseph Priestley studied the phlogistic bubbles generated when malt fermented into beer, then decided to try transferring them to water under pressure. Perhaps in quest of a nonalcoholic brew for his sodden parishioners, he quafed this seminal soda pop and started a fad that roared on through the age of quack tonics to Casey Jones's favorite mouthwash (alas, no more cocaine in Coke after 1903) to the Pepsi Generation. The quintessential greenhouse drink: Joe Cola, in honor of the estimable Priestley.

• carbon monoxide •

The gas that made the phrase "colorless, odorless, tasteless" something of a spine tingler. It is also flammable, but this characteristic does not figure in classic sealed-garage scenarios.

If we inhale enough carbon monoxide, it will kill us because it bonds strongly to hemoglobin molecules in our blood, displacing oxygen vital to the brain and heart. (Hemoglobin's affinity for CO is about 240 times greater than for O_2.) Evidently, you don't have to try hard: between 1979 and 1988, carbon monoxide poisoning caused 11,547 *unintentional* deaths in the United States, 57 percent of which were due to motor-vehicle exhaust. No other single substance causes more fatalities.

Carbon monoxide is produced at myriad sites of combustion (i.e., burning of organic fuels), including gasoline engines, natural and man-made fires, stoves, ovens, space heaters, water heaters, furnaces, charcoal grills, and tobacco smoking. Paint strippers containing methylene chloride give off fumes that our body metabolizes into CO, hence the warning on the can for generous ventilation. Cigarette smoke contains about 1 percent CO by volume (or 10,000 parts per million)—another reason to regard fags as portable suicide systems.

Federal air-quality standards dictate that CO concentrations not exceed 9 ppm for an eight-hour average or 35 ppm for one hour, though these thresholds are frequently broken in many American cities. The best way to relate dosage to biological effect is to measure the concentration of hemoglobin glommed by carbon monoxide (COHb in chemical notation) as a percentage of total available hemoglobin. Thus, nonsmokers normally have COHb concentrations in their blood of around 0.5 percent. Smokers zoom much higher, from a median of 5 percent to a maximum of over 15 percent. Such figures do not always correlate with either the clinical condition of or prognosis for a severe poisoning victim, however.

Overexposure, starting between 15 and 25 percent, can cause headaches, dizziness, confusion, weakness, nausea, vomiting, and loss of muscle control. The best first aid is lots of fresh air; breathing normal air will clear away half of blood carbon monoxide in about five hours. Prolonged exposure leads to unconsciousness, brain damage, and death. Even long-term, low-level intake may bring persistent memory loss and muscle pain. On Manhattan's Upper East Side in February 1997, a clogged chimney trapped CO from a natural gas furnace inside an apartment house, leading to one death and nine hospitalizations. Paramedics and doctors at first mistook headaches and shortness of breath among residents for influenza.

At blood concentrations above 40 percent, victims need immediate treatment with pressurized pure oxygen, known as hyberbaric oxygen therapy. Levels over 60 percent are usually fatal. Experiments have shown that healthy adults start losing aerobic exercise capacity, visual perception, vigilance, and manual dexterity at around 5 percent, confirming the *Via recta*'s warning of 1620: "Tobacco drieth the brain, dimmeth the sight, and causeth the blood to be adjusted."

People with chronic heart or lung diseases are obviously more sensitive. And, yes, those bridge and tunnel workers who collect your toll are at much higher risk for heart attacks than the general population. Video freaks, too, beware; a 1984 study found significant decrements in game performance at just 2 percent COHb.

As if CO were not bad enough by itself, military scientists long ago figured out how to make it worse. Phosgene, which is produced by combining carbon monoxide with chlorine, became one of the first weapons of modern chemical warfare. (It was developed by Fritz Haber—who won a Nobel Prize for discovering how to synthesize AMMONIA—and introduced by the German Fifth Army at Verdun, France, in 1916.) Phosgene is considered lethal when mingled with 9,999 parts of normal air (100 ppm), making it twice as deadly as chlorine alone.

• **comet tails** •

In Greek, they were called *aster kometes,* or long-haired stars. Aristotle believed that comets dwelled high in the Earth's atmosphere, just like the Milky Way. The Sun drew from the planet a warm "exhalation" that sometimes caught fire. A hundred years later, the mathematician Apollonius made a better guess, opining that comets were celestial objects that became visible when they reached the lowest part of their course, which in a geocentric universe was here. Neither of these two geniuses possessed any evidence other than what even the hoi polloi could see with their own eyes from time to time.

As for why comet tails always pointed away from the sun, no one dared say, the religious implications being unspeakable. Science did not generate a worthy theory until 1900, when the Swedish chemist Svante Arrhenius proposed that material in the tail was deflected by solar radiation. Comets are now known to possess two tails, one made of plasmatic gas molecules (usually ionized CARBON MONOXIDE and nitrogen) and the other of DUST particles. The first type is pushed askew by "solar wind" protons and electrons streaming out of the Sun at 892,800 miles per hour, the second by PHOTONS of sunlight. The fact that they both point sideways to the comet's path

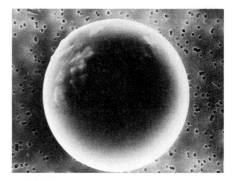

Comet dust particle collected from the atmosphere.

when it gets close to the Sun raises the possibility that the Earth might pass through them without actually hitting the comet's iceball nucleus.

Indeed, the Earth probably careened through the tail of Halley's comet on May 19, 1910. Newsstands sold a lot of papers that day by trumpeting the poisonous constituents of cometary gas, such as HYDROGEN SULFIDE (H_2S), cyanogen (CN), hydrogen cyanide (HCN), and methyl cyanide (CH_3CN), without explaining that their density was so low that they would burn out in the upper atmosphere. But micrometer-sized comet dust grains containing nearly a score of basic elements (in decreasing abundance: hydrogen, oxygen, carbon, silicon, magnesium, sulfur, iron, nitrogen, sodium, aluminum, calcium, nickel, chromium, manganese, titanium, cobalt, potassium), as well as organic compounds that might have played a role in the origins of life, are slowed by atmospheric drag and thus accumulate. They have even been collected by U-2 aircraft. Though the density of a comet's dust tail is much less than cigarette smoke, it extends for millions of miles. Comets may therefore have transported much, if not all, of the biogenic organic material (such as METHANE) from the outer solar system, where it is common, to the inner planets, which lacked it primordially but basked in a water–based chemistry conducive to life.

• cooties •

A cutesy schoolboy epithet that traces back to the decidedly uncute conditions of trench warfare that many schoolboys found themselves in during World War I. It probably derives from *kutu* in the Malay language of colonial regiments. By now, most taunting kids never realize it refers to body lice, *Pediculus humanus,* which will infest anybody without access to

clean clothes, soap and water, or other rudiments of personal hygiene.

Besides being an itchy nuisance, body lice are potentially dangerous because they may carry germs that cause typhus, trench fever, and other diseases. In typhus epidemics, the lice feed on an infected individual's blood, jump to another host, and defecate *Rickettsia prowazekii* microbes, which get scratched into the skin. A week or two later, the victim develops a prolonged fever of 104° F and a rash over the entire body. If the condition is allowed to progress, derangement and organ failure commence, with fatality rates of around 60 percent in older, untreated victims. Antibiotics can defeat typhus if used early on, but the lice must be eliminated to break the chain of infection, usually by dusting with insecticides.

A milder form of typhus, known as murine typhus but sometimes called "Urban Typhus of Malaya" (hence the linguistic readiness of those colonial troops), is transmitted to humans by a different critter, the rat flea (*Xenopsylla cheopsis*). To sick and dying soldiers in the Western Front's squalid trenches, of course, cooties were anything that itched. Besides, even medical scientists of that era did not fully understand the link between such pests and disease; two pioneering researchers, Howard T. Ricketts and Stanislas von Prowazek, both died of typhus.

Several breeds of *P. humanus* are also responsible for relatively benign infestations. *P. humanus capitis* prefers the head and loves to attend school regardless of tuition level. *P. humanus corporis* resides in dirty underwear, from where it ventures out onto the bottom and tummy (no wonder the kids have been giggling for so many generations). *Phthirus pubis,* a close cousin affectionately known as the crab louse, is usually transmitted by intimate sexual contact. All these cooties are susceptible to medicinal shampoos and diligent laundering. *Nyah, nyah, pants on fire!*

• **cosmic rays** •

Whenever the Apollo astronauts closed their eyes to stack some *z*s while cruising to and from the Moon, they were startled by pinpoint flashes of light. Nature's alarm clock? *Don't fall asleep at the wheel out here, bub!* No, just atomic nuclei from the far reaches of the cosmos, striking their retinas to remind them they weren't in Kansas anymore.

In 1925, the American physicist Robert Millikan (1868–1953) named this type of radiation cosmic rays, thereby proving himself a match for any Hollywood sci-fi writer. Unfortunately, he also tried to take credit for discovering the phenomenon, which had already been done by someone else in 1912. Victor Hess (1883–1964), an Austrian physicist, secured the 1936 Nobel Prize not just by beating out Millikan, but by having demonstrated more pluck than a French Jesuit. In 1910, Father Theodor Wulf had climbed the Eiffel Tower with radiation detectors, recording higher numbers at the top and speculating that the source might therefore be heavenward. He thought about taking his meters up in a balloon, but apparently couldn't stomach heights loftier than Monsieur Eiffel's observation deck. The hardier Hess soon took the trip, finding that the radiation intensity several miles high was three to five times greater than on the ground. Measurements taken during daytime were the same as those at night, so the Sun was obviously not the source. *Cosmic* rays, indeed, though nobody knew what they were composed of.

Even in the atmosphere's upper reaches, cosmic rays are so thinly spread that Hess's primitive instruments could only record their existence, not identify their content. It took the advent of the Geiger counter in the late 1920s for detecting charged particles, refinement of the cloud chamber (which had been invented in the late 1890s) for showing their tracks, and development of special photographic emulsions after World War II to provide the level of sophistication needed to determine just what the mysterious rays were: interstellar detritus, namely 86 percent protons, 12 percent helium nuclei, and the rest carbon, oxygen, and rarer nuclei, plus a few electrons and positrons. Further research cataloged the subatomic avalanche, or "air showers," created by collisions with atmospheric elements on the way down, where the dissipated energy is sometimes visible as aurora borealis.

Cosmic rays come from two sources: stars throughout the universe, many so far away that particles reaching us today started before Earth existed; and the Sun, which sends out bursts of radiation during sunspot activity and solar flares. In 1949, Enrico Fermi theorized that cosmic rays get their energy from shock waves generated by exploding stars, or supernovas. The Van Allen radiation belts that surround Earth along its magnetic field trap vast numbers of the particles arriving from outer space. Others are blocked by the atmosphere. Without these barriers, cosmic radiation would be intense enough to make life impossible.

For every rise of 10,000 feet from the Earth's surface, the penetration of cosmic rays is three times greater. Thus, the annual flight time of airline crews who routinely work at 30,000 feet and above is limited by government regulation to avoid dangerous levels of radiation. A single transatlantic flight may result in whole-body exposures a bit greater than those received in a whole-mouth dental X-ray series. Exposure on the surface varies with location, with intensities greater near the poles. Tropical areas therefore receive minimal radiation, with Scandinavia, Canada, and other high-latitude regions getting up to three times more, depending on altitude. In the United States, Colorado and Wyoming register the highest exposures, Georgia and Louisiana the lowest.

Before man-made accelerators, or "atom smashers," appeared in the 1950s, cosmic ray research produced the first evidence of a veritable zoo of "fundamental" particles beyond the simple proton-neutron-electron model of the atom. With the demise of the multibillion-dollar Superconducting Super Collider project in 1993, the golden age of particle accelerators may finally have ended. It was designed to impart energies of about 40 trillion electron volts (TeV, or 10^{12} eV), where one electron volt equals the energy an electron gathers by traveling across the terminals of a one-volt battery. For comparison, lighting a match releases about 10 billion TeV, but the energy is widely distributed compared with that of a concentrated beam of particles in an accelerator.

The most powerful existing accelerator, Fermilab's Tevatron, operates at 2 TeV. Cosmic ray energies can reach upward of 10^{21} eV, far outclassing any feasible man-made device. But these biggies are rare. A detector array known as "Fly's Eye" in the Dugway desert of northwestern Utah has counted only several hundred particles above 10^{10} giga, or billion, electron volts (GeV) in a decade of operation. Below about 10^9 GeV, galactic cosmic ray particles are deflected by magnetic fields in the Milky Way, which means they arrive at the Earth from random directions rather than on a beeline from their source. Above about 10^{11} GeV, the major constituents of cosmic waves, protons, slow down because of an interaction with the universe's BACKGROUND RADIATION, lowering the likelihood that any will reach us. The few ultra-high-energy rays detected by Fly's Eye probably come from beyond the Milky Way, since they are too strong to be contained by our galaxy's magnetic fields.

Two well-known radionuclides that we are exposed to constantly—tritium and carbon-14—have been generated for millions of years by cosmic rays. Most of the tritium is in ocean water and results in extremely low doses. As for carbon-14, every compound in the human body contains it, so find something else to worry about.

• dioxin •

The shorthand name for seventy-five different compounds made of carbon, oxygen, hydrogen, and chlorine atoms. Only seven are highly toxic, but one of them is a doozy.

Dioxins are totally useless and have never been manufactured on purpose. They are notorious pollutants, an unwanted by-product of virtually any industrial chemical process involving chlorine. Until the advent of large-scale chlorine chemistry between the world wars, dioxins existed only at insignificant levels. Today they are as ubiquitous as the 90 billion pounds of chlorine produced every year worldwide for commodities such as plastics, paper, pesticides, and solvents. Because they "bioaccumulate," everyone in the industrialized world carries some in their body, rather like the plutonium from atomic bomb FALLOUT.

All dioxins are formed chemically by the linkage of two benzene molecules, each of which contains six carbon atoms and six hydrogen atoms in a distinctive hexagonal ring. The geometry and oxygen content of the hookup, as well as the number and location of chlorine atoms substituted for hydrogens, determine which of the various dioxins is obtained. Dioxinlike compounds called furans and biphenyls (which comprise the equally notorious PCBs) also cook this way. The size and shape of the final molecule, which are fixed by the total number and positions of the chlorine atoms, decide its toxicity. The most heinous member of the dioxin clan is thus designated in the jargon of organic chemistry as 2,3,7,8-tetrachlorodibenzodioxin (mercifully abbreviated TCDD), where the numbers tell the location of four (*tetra*) chlorines (*chloro*) on two benzene rings (*dibenzo*) linked by two oxygens (*dioxins*). At room temperature, it is a colorless crystalline solid.

When the word *dioxin* appears in the popular media, it usually means this supremely poisonous compound, even though TCDD is rarely found in the environment by itself.

Just how bad is it? In the 1970s, experiments with rats found greatly increased cancer incidence at feed exposures as low as 5 parts per *trillion* and, in an important study sponsored by Dow Chemical Company, at 210 ppt. These figures meant that dioxin was the most potent animal carcinogen ever tested. Since then, numerous trials have proved that TCDD causes tumors in multiple species of both sexes at many different body sites. Indeed, no "safe" low dose has ever been established. Animal studies have also demonstrated that dioxin kills fetuses, causes birth defects, and suppresses the immune system at infinitesimally small doses. Rhesus monkeys, for example, have suffered increased prenatal mortality at daily doses of 0.64 nanograms per kilogram of body weight. (A nanogram is *one-billionth* of a gram.) "Acceptable" daily human doses, a concept influenced perhaps as much by politics as by science, range from the U.S. Environmental Protection Agency's 0.006 picograms (*one-trillionth* of a gram) per kilogram of body weight per day to the World Health Organization's 10 pg/kg/day. Hence the mantra that dioxin is "the most toxic chemical known to man."

Dioxin that leaks into nature tends to stay there because it does not dissolve well in water or evaporate much into the air, sticks to organic matter, and rarely breaks down into other substances. To boot, it is very soluble in oils and fat, which means that if we eat some, we don't soon excrete it but store it in fatty tissue. The EPA estimates that people consume on average 119 pg/day (eggs, 4 pg; fish, 8 pg; pork, 12 pg; chicken, 13 pg; milk, 17 pg; dairy, 24 pg; beef, 37 pg; fruits and vegetables, negligible), which amounts to more than 280 times the "acceptable" dose. The daily intake from breast milk is 10 times higher—difficult news for the La Leche League, but not a good enough reason to switch to formula. Dioxin accumulates until a middle-aged person carries an average 9 ng/kg "body burden," which is just short of the lowest level known to harm human health (decreased testes size, altered glucose tolerance). In other words, if there is a safe threshold, we're all near or beyond it.

For scientific and ethical reasons, "acceptable" human dose figures rely primarily on animal research. Only one study, a dubious exercise directed by Dow in a Pennsylvania prison in 1965, has ever exposed people deliberately to TCDD. (Subjects were evaluated for a severe skin disease called chloracne, which is the classic symptom of acute exposure.) Epidemiologic evidence for dioxin as a human carcinogen comes from studies of chemical plant accidents, broadside contamination such as that from Agent Orange herbicide during the Vietnam War, and long-term occupational exposures. Conclusions are mixed, with much ongoing debate about levels of exposure, the influence of other chemicals, and even the honesty of the research. The EPA and the International Agency for Research on Cancer (IARC) rate TCDD as a probable human carcinogen, associated with tumors in soft or connective tissue, the lungs, liver, stomach, and with non-Hodgkin's lymphoma. A devilish list of male and female reproductive problems, fetal disorders, nervous system damage, immune system glitches, and metabolic/hormonal changes round out the portrait of a public health menace.

Because dioxin pollution is entwined with powerful industries basic to modern life, it is rivaled perhaps only by nicotine in the politicization of science. Monsanto, BASF, Dow, and Hoffmann-La Roche are among the corporations whose products have been implicated in famous dioxin disasters, from the 1949 explosion of a Monsanto herbicide plant in West Virginia, to a 1953 release at BASF's trichlorophenol (a starting material for 2,4,5-T—the herbicidal partner of 2,4-D in Agent Orange—and the antibacterial agent hexachlorophene) factory in Germany, to Dow's military production of Agent Orange in the 1960s, to a runaway reaction at Hoffmann-La Roche's chemical works in Italy in 1976 that released six pounds of TCDD over 275 urban acres. Meanwhile, the catastrophic negligence of Hooker Chemical Company at Love Canal, New York, and waste oil handlers around Times Beach, Missouri, attracted press coverage that turned dioxin into a household word. Because TCDD is dangerous to human health in such minute quantities, which can be extraordinarily expensive to clean up from contaminated soil or water, government regulatory policy continues to be a vicious battleground. Serious laypeople are therefore advised to spend plenty of time in the library with primary sources.

Anyone living near a past or present waste incinerator; paper mill, metal smelter, wood treatment site, petroleum refinery, cement kiln, or plant that makes plastics, solvents, or pesticides involving chlorine; or heavily trafficked roads is at risk of elevated dioxin exposure. Airborne ash from incinerators spreads dioxin thousands of miles from point sources, the primary reason for its ubiquity. Oh yes, it's in cigarette smoke, too.

• dust •

Awareness of house dust, on the linguistic if not maidservant level, is as old as civilization. A prehistoric Indo-European base, *dheu-,* evolved into the Sanskrit *dhuma,* meaning "smoke," which Proto-Germanic speakers adopted as *dunstu-,* for a cloud of fine particles, which was the basis for Old English *dust* appearing around A.D. 725. Northern European abodes of

Dust, disease, and social status.

every social stratum doubtlessly harbored great balls of it, as did the unwashed, heavily garbed inhabitants. Dust has long been abhorred by polite society, but although it can be easily vacuumed or mopped or wiped out of sight, it is still everywhere. Go away for a few days and your home looks like Morticia Adams's.

House dust, as opposed to the technical category of PARTICULATES, consists of the detritus of human domestic life: food crumbs, fibers, skin flakes, grit, plant matter such as POLLEN and SPORES, microbes, MITES, FUNGI, and debris from insects and domestic animals. Many of these ingredients contain ALLERGENS, but enough dust will make anyone sneeze. Though dust can sometimes be directly toxic, its association with diseases such as asthma, dermatitis, and rhinitis applies only to hypersensitive individuals.

Doctors used to blame dust allergies on pet dander (hair with skin scrapings, like dandruff), insects, and fungi. But research since the late 1960s has shown that dust mites of the family Pyroglyphidae, especially the genus *Dermatophagiodes*—which thrive on skin scales in carpets, stuffed furniture, and mattresses at just the temperature range humans prefer, 65° to 80° F—are the primary culprits in most parts of the world. They especially enjoy relative humidities above 45 percent. Diligent scientists found that homes of symptomatic people have more than 500 mites per gram of dust. Exceptions occur in high-altitude or dry climates, which the mites don't relish, as well as city neighborhoods, where rodent urine and roach debris take charge.

Most of the mites in dust are dead. Their little ca-ca balls, pardon the old expression, contain the allergens. About the diameter of pollen (10–35 microns), each carries around 0.2 nanograms (billionths of a gram) of allergenic material, multiplied by the astronomical factors of even the most Swiss household. If you could arrange a mite holocaust, as happens naturally during periods of low humidity, it would still take months for indoor allergen levels to decrease.

Besides mites, common biological sources of allergens in house dust include spiders, crickets, flies, locusts, beetles, fleas, moths, midges, and cockroaches; cats, dogs, ferrets, skunks, horses, rabbits, pigs, mice, rats, gerbils, and guinea pigs; kapok; and *Penicillium, Aspergillus, Rhizopus,* and *Cladosporium* fungi. Roach allergens have been closely studied, due to their urban ubiquity, and are believed to come from the pests' feces and saliva.

A special word about cats, which are powerful asthma triggers. There is no such thing as a nonallergenic cat (or dog, for that matter). Licking and grooming spread the primary allergens, which are then found in settled dust *and* constantly airborne in the homes of pet owners. Even many houses that are cat-deprived harbor high levels of contaminated dust brought inside on the clothes of cat lovers. Dust particles carrying allergen are often small enough (less than 2.5 microns in diameter) to be inhaled deeply into the lungs. Beds and furniture where the little darlings sleep are vast storage sites of allergens. If an offending cat is banished from the house, it takes at least five months for the itchy dust level to reach that of a no-cat home.

Attention unemployed Ph.D.s: Pharmaceutical companies manufacture dust extracts for allergy skin tests from the contents of home vacuum cleaner bags. No pets, please. Call now!

• **EMF** •

In 1979, researchers Nancy Wertheimer and Ed Leeper published an article in the *American Journal of Epidemiology* titled

"Electrical Wiring Configurations and Childhood Cancer."
They presented evidence, inconclusive but untrivial, that
otherwise extremely rare cases of leukemia were turning up
among Denver children living near power transmission lines
and transformers. It was the sort of report that rattles an
advanced capitalistic society to its bones.

In 1996, a blue-ribbon National Research Council committee mandated by Congress to get to the bottom of the issue
announced that there is "no conclusive and consistent evidence" that ordinary exposure to electromagnetic fields (EMF)
in the home can "produce cancer, adverse neurobehavioral
effects, or reproductive and developmental defects."

That is the long and short of it. In between, and still running, are more than 500 studies of whether EMF causes myriad health problems, such as brain and breast cancer in electrical
workers, miscarriages among computer users, and leukemia
and behavioral problems in children. In the leukemia cases, an
apparent increase was found around homes within fifty yards of
high-capacity (115,000–500,000 volts) transmission lines—the
kind with six thick wires typically strung between steel towers.
Yet when EMF measurements are taken in the homes of sick
children, no correlation is obtained between field strength and
cancer incidence.

Because childhood leukemia is so rare—appearing in only
about 4 out of every 100,000 American kids between the ages
of two and ten—even the best studies have just a few dozen
cases to work with every year, which severely hampers drawing
wider conclusions from the data. In one of the few large-scale
studies available, 400,000 adults in Finland were scrutinized for
residential EMF exposure over a twenty-year period. No
related cancer risk emerged. In 1997, the National Cancer
Institute published a seven-year study of 1,000 children in the
New England Journal of Medicine, accompanied by an editorial

that bluntly said "it is time to stop wasting our research resources" on looking for a connection between EMF and leukemia.

Science can never prove that anything is safe, of course. It can only say that hazards have not been found. In this case, the picture is clouded by the fact that empirical evidence remains unsupported by theory. Household-level EMF has never been linked to adverse health effects in the laboratory. There is no accepted biophysical mechanism for EMF carcinogenesis. If there is a causal effect at work, the NRC suggested, it may have nothing to do with EMF, but rather reflect such known risks as living near automobile pollution or in lower socioeconomic conditions.

Nonetheless, millions of dollars have already been spent to reduce EMF levels in residential areas. Giant financial interests in the cellular phone and electrical power industries, which fund many EMF research projects, joust perennially with academic epidemiologists over how to interpret the cloudy data. If a process by which low levels of nonionizing radiation can damage living tissue were ever discovered, stronger population studies could be designed. Meanwhile, the threat—real or imagined—has become part of the endless battle among developers, utility companies, real estate agents, property owners, and environmentalists.

To understand what EMF *is,* however, requires a short physics lesson. The electromagnetic force is one of three fundamental forces that govern the behavior of atoms. (The others are the "weak force," which causes radioactivity, and the "strong force," which binds the nucleus together. A fourth force, GRAVITY, holds the universe together but has no measurable action within individual atoms.) Electromagnetism keeps the negatively charged electrons in their "orbits" around the positively charged nucleus, making atoms and bulk matter seem solid. The theory of electromagnetism was published by the Scottish physicist James Clerk Maxwell (1831–1879) in 1865 and has evolved into the modern theory of quantum electrodynamics, which is mercifully beyond the scope of this modest discussion.

When electrons flow through a wire in an electrical current, they create a magnetic force around the wire that can start currents flowing in other conductors, including living tissue. The strength of this magnetic "field" decreases by the square of the distance from the wire. At 50 feet from a 230,000-volt power line, the field is about 20 milli-gauss (mG, or thousandths of a gauss, a unit of field strength). At 200 feet away, it's 1.8 mG. Homes and offices expose people to fields of about 0.1 to 3 mG. By comparison, the Earth's magnetic field is 500 mG. Electrical currents induced in living tissue by household EMF are a thousand times weaker than the nerve signals produced naturally in the brain.

Even at levels a hundred times higher than in an average home, there is no evidence of adverse health effects. In fields 1,000 to 100,000 times higher, cultured cells show changes, but none has been connected to illness. One such alteration is even known to be beneficial: pulsing fields of just 5 mG help animal bones heal, a phenomenon that has found practical use in veterinary care.

Residential electrical currents are called "alternating" (AC) because they change direction sixty times every second. At much higher frequencies, especially in the MICROWAVE range of 0.3 to 300 billion cycles per second, EMF can heat up exposed biological tissue (hence, microwave cooking). Skin, eyes, and testicles have long been known to be sensitive to such thermal effects. For thirty years, governments have tried to set occupational exposure guidelines, at first mostly with the communications industry and military in mind, but increasingly for the workaday environment of anyone who spends time around electronic devices that radiate EMF (computers, TVs, cellular phones, microwave ovens, diathermic medical instruments, etc.). Because bioelectromagnetic research has always lagged behind the marketing of commercial products, these standards tend to be both arbitrary and contradictory, especially from one country to the next.

Should you buy a house near those power lines? If it's all you can afford, it beats sleeping outside in the rain.

• fallout •

"Gee, that's important, isn't it?" President Eisenhower said after his first briefing on radioactive fallout. "Is everybody ready for lunch?"

The entire government was out to lunch for most of the 1950s when it came to fallout. The long-range consequences of nuclear weapons, as opposed to the ground-zero horrors of Hiroshima, were essentially unknown to the public until March 1, 1954, when an H-bomb code-named Bravo exploded seven feet above Namu Island in the once paradisial Bikini atoll. Calculated beforehand to be equivalent to 8 million tons of TNT, Bravo popped off at fifteen megatons instead. Namu disappeared in a titanic mushroom cloud of vaporized coral, wafting back down a few hours later as white radioactive powder over an area of 7,000 square miles.

Not only did the physicists predict the wrong yield, but meteorologists misjudged the wind. The ash therefore fell on Marshall Islanders and American bomb testers instead of empty ocean. The islands of Rongelap, Rongerik, Alingnae, and Utirik (which was 300 miles from Bikini) were evacuated, but not before the mostly barefoot, tropically attired (i.e., nearly naked) Marshallese suffered disfiguring radiation burns and body doses that would undermine their health for the rest of their lives. Two-thirds of Rongelapers developed nausea within fifteen hours; half lost their hair after two weeks, indicating hellish exposure.

At first, American authorities covered up the disaster. Roger Strauss, chairman of the Atomic Energy Commission, declared that the islanders were "well and happy." But they did not realize that a Japanese fishing boat named *Fukuryu Maru,* the *Lucky Dragon,* got caked in fallout while motoring eighty-five miles east of Bikini, well outside the test zone. Two weeks after Bravo, her crew returned home with all the classic symptoms of radiation sickness rather well known in Japan: nausea, fever, internal bleeding, skin burns. To boot, her catch was found to be hot, setting off a fish scare in a country where seafood is a staple of life. On September 24, the *Lucky Dragon*'s radioman succumbed in Tokyo, thus becoming the first atomic fatality since Nagasaki. Washington apologized and sent his widow $2,800.

The cat was out of the bag, however, about fallout, which helped stoke the era's mass paranoia. Such darksome nightmares as the "cobalt bomb," an H-bomb specially coated to maximize release of cobalt-60 around the globe, entered political rhetoric and schoolboy vernacular. Adlai Stevenson called for an international test ban during the 1956 presidential campaign, warning that "poisoning of the atmosphere" was threatening "the actual survival of the human race." But Republicans were able to portray him as naive and dangerous to national security—a charge that would continue to work like a charm against all critics of the military for decades.

Meanwhile, American, Soviet, and British weapons designers detonated scores of nukes in the open. Having worn out their welcome in the South Pacific, U.S. forces moved to the

Nevada desert eighty miles from Las Vegas. The AEC, for its part, never waivered in assuring the public that fallout from these tests—even the *Russian* ones—was no hazard whatsoever. Gradually, technical facts to the contrary salted the debate thanks to individuals like Nobel laureate Linus Pauling and organizations like SANE, but mainstream press organs such as *Time, Newsweek,* and *U.S. News and World Report* routinely smeared them. "Much of the clamor about the danger of fallout is inspired by Communists," declared *U.S. News* in a typical article.

Not that officials were scientifically ignorant, of course. After Trinity, the premier A-bomb test in 1945, fallout over the Midwest had contaminated Kodak packaging material, which in turn fogged lots of unsold film (harking back to Henri Becquerel's discovery of radioactivity in 1896; see ALPHA RAYS). When it happened again in 1951, Kodak threatened to sue the government, but the AEC promised to start giving the company warning of Nevada tests. Informing the public, however, always fell by the wayside when faced with "the more immediate and infinitely greater dangers of defeat" by the Soviet Union, as Roger Strauss intoned. Instead, American citizens received helpful tips such as the army quartermaster general's 1958 suggestion that eating "cabbage and broccoli may be the means of doubling the capacity of man to withstand" fallout.

Edward Teller, father of the H-bomb and perennial advocate of anything nuclear, spearheaded the effort to soothe public fears about fallout. "Our custom of dressing men in trousers causes at least a hundred times as many mutations as present fallout levels," he said. "But alarmists who say that continued nuclear testing will affect unborn generations have not allowed their concern to urge men into kilts."

In 1957, Teller and others even managed to convince Eisenhower of the imminent development of a "clean bomb" that would cause 96 percent less fallout than old-fashioned dirty ones, if only atmospheric testing could continue unabated. But oh, that 4 percent! This was the same concept that would later be retooled as the "neutron bomb," a relatively low-yield H-bomb detonated at high altitudes. Ike swallowed it until Nikita Khrushchev told him he was talking "stupidities." Even a pure fusion weapon exploded in air, with no fission trigger, would still create fallout of carbon-14, tritium, and other nuclides, as Dr. Teller knew perfectly well.

Somehow, against great odds, the public grasped the truth, as shown by a 1957 Gallup poll that found 63 percent of Americans favoring a test ban treaty. The blockbuster novel *On the Beach,* about a world terminally poisoned by fallout, perhaps outdid all the federal propaganda. In 1958, the Big Three nations (the United States, USSR, and Great Britain) agreed on a moratorium. In 1959, strontium-90 contamination of milk struck at the fearful hearts of families regardless of political affiliation. In 1961, President Kennedy matched the Soviet resumption of weapons tests but acknowledged the rising pressure of public opinion by stipulating they would take place only "underground, with no fallout." Finally, in 1963, the Big Three ratified a ban on all atmospheric tests, as well as those undersea and in space. It was a rare diplomatic success in the long Cold War.

What relevance does fallout have today? Besides the fact that other nuclear nations—China, France, India, Pakistan, South Africa, Israel—never signed the treaty and have tested bombs openly or secretly since 1963, contamination from those eighteen hectic years after August 1945 is still with us. By the late 1960s, more than half of the external radiation dose from these tests had been delivered to the world's population, plus two-thirds of that from internally deposited cesium-137. But even by the year 2000, only one-tenth of the internal dose from carbon-14 will have been delivered. Though it accounts for a small portion (less than 0.3 percent) of our total annual exposure from all sources (see BACKGROUND RADIATION), fallout is the reason almost every person on Earth carries a bit of plutonium in their bodies that nature obviously did not intend. Cesium-137, with a half-life of thirty years, remains a source of external GAMMA RAYS.

People who lived directly downwind from American tests, especially in southwest Utah, have been financially compensated by the government. But a recent National Cancer Institute study of fallout patterns from 1951 to 1958 in all forty-eight contiguous states found that some of the highest exposures occurred in Albany, New York, and parts of Massachusetts. All 160 million U.S. residents of that era received some fallout, with the average national thyroid dose around 2 rads. People in twenty-four counties with the greatest levels may have received an average of 16 rads. Children aged three months to five years living in those hot spots might have absorbed as much as 160 rads from milk contaminated with iodine-131. The NCI estimated that 10,000 to 75,000 Americans may develop thyroid cancer because of bomb fallout—an increase of about 20 percent over normal incidence.

The medical consequences are subtle on a macro level, but all too real. Fallout from the era of atmospheric testing peaked in 1957 and 1962. Death rates for acute and myeloid leukemia in American children five to nine years old also peaked in 1962, and again in 1968. Leukemia mortality for all ages and cell types peaked in that same decade and was highest in states with high levels of strontium-90 in diet, milk, and bones and lowest in states with low ^{90}Sr. Among residents of Utah and neighboring states downwind of the Nevada test site, as well as among military personnel who participated directly in the tests, studies have found excess cancer, primarily leukemia. The same is true for British participants in weapons tests in Australia and the Pacific Ocean.

And then there is the ongoing development of new nuclear weapons, such as the B-61-11 "bunker buster" designed to destroy underground structures. Actually an old H-bomb housed in a new shell made of depleted uranium (which is radioactive itself), it would be dropped without a parachute and burrow into soil before detonating. The B-61-11 replaces a Bravo-magnitude bomb, which would have accomplished the same mission by creating a 500-foot-deep crater, while also spawning a fireball two miles in diameter and colossal amounts of fallout. The new device sports a handy selectable yield from 300 tons to about 340 kilotons, which would cause less "collateral damage," as the Pentagon techies call death of civilians. (For comparison, the Hiroshima atomic bomb was 22 kilotons, and the conventional Oklahoma City terrorist bomb was somewhat under 2 tons.) Of course, any nuclear weapon that kicks up dirt makes lots of fallout. If the B-61-11 were ever used to take out an underground Libyan or Iraqi munitions factory, for example, vast sections of the Middle East would be hot in more than the Mad-Dogs-and-Englishmen sense.

The most contemporary use of the term *fallout* has nothing to do with radiation, however. Toxic chemical fallout containing DIOXIN emitted from more than a thousand trash-burning incinerators in the United States and Canada spreads contaminated DUST across the entire continent. Half of the dioxin in the Great Lakes thus comes from as far away as Texas and Florida, according to a 1995 study by the Center for the Biology of Natural Systems at Queens College-CUNY, based on National Oceanographic and Atmospheric Administration computer models and actual air samples. The Environmental Protection Agency has found that ingestion of dioxin that gets into the food chain is the primary source of human exposure.

• flatus •

Farts.

More than 400 species of BACTERIA inhabit the human gut in a continuous chemical stew. We should be glad, because they help digest food and synthesize substances, such as vitamin B, that we very much need. There are nefarious methanogens among them, however, that produce METHANE as they go about their business, as well as gassers like *Clostridia, E. coli, Bacteroides,* plus various yeasts and protozoans. According to a courageous bit of research published in 1976 by Messieurs Levitt, Lasser, Schwartz, and Bond with the epic title "Studies of a Flatulent Patient," the resultant wind that blows from our bottoms is composed mostly of nitrogen, oxygen, hydrogen, methane, and CARBON DIOXIDE in varying percentages depending on what we've been eating. Indigestible carbohydrates are prodigious sources of H_2, thus the antisocial reputations of beans, corn, and potatoes. Sometimes sulphate-reducing microbes get in there, too, and squeeze off HYDROGEN SULFIDE of rotten-egg infamy, which fosters the beneficial effect of lowered methane volume, but the social disaster of cheesy *odeur.* Hence those hot little poohs that can clear a station wagon.

Quantity and frequency are the stuff of family legend. A single meal could theoretically pump as much as four liters of carbon dioxide into Uncle Irv's duodenum, but with luck much of it will be absorbed into the blood. One study of men between 25 and 35 counted an average of 13 ± 4 farts a day, but there are individual cases on record at the 140 level. And then there was the Moulin Rouge musician known as "Le Petomane" (perhaps sharing a root with *pétoire,* popgun), whose instrument was his backside.

Few words in English possess more ancient cousinage throughout the Indo-European languages than "fart," which traces back to prehistoric *perd-*, probably an onomatopoeic creation. German *farzen*, Swedish *fjarta*, Danish *fjerte*, Russian *perdet*, Polish *pierdziec*, Greek *pordizo*, and Welsh *rhechain* all share this powerful ancestor. Old Norse *fisa*, however, took a more musical turn, emanating from the Indo-European base *peis-*, the sound of blowing or breathing out, which also generated Slavonic *piskati*, whistle, and Serbo-Croatian *pistati*, hiss.

• formaldehyde •

A colorless gas at room temperature that is known to biology students and horror movie fans for the pungent-smelling solution in water used to preserve animal specimens. It occurs naturally in the environment and the human body in minute quantities, but its status as one of the top ten industrial organic chemicals, as well as its presence in automobile exhaust, tobacco smoke, and building materials, has made it one of the twentieth century's ubiquitous health hazards.

Formaldehyde was first prepared in 1859 by a Russian chemist, Alexander Mikhailovich Butlerov, who made synthetic sugars from it. (Butlerov pioneered the use of formulas to represent the position of atoms in compounds and coined the term *chemical structure*.) In 1889, France and Germany granted the first patents for commercial production. Initially used in the manufacture of dyes, it soon became valuable as a disinfectant and embalming agent, which were its chief applications in the United States at the turn of the century. The development of synthetic resins caused output to rise steeply, fueled by war demands. During World War II, the U.S. government built two formaldehyde plants, at Danville, Pennsylvania, and Morgantown, West Virginia, to support production of cyclonite, or RDX, a military explosive. After the war, the booming plastics industry took up the supply. Today, over 5 billion pounds are used annually in the United States alone in chemical synthesis and products such as polyester clothing, permanent-press fabrics, photographic film, cosmetics, furniture, foam insulation, plywood, fertilizers, germicides, paints, particleboard, Formica, carpets, and wall paneling. It is impossible to escape man-made formaldehyde in the developed world.

In older chemistry books, formaldehyde was notated as CH_2O, but the suggestion of an inherent water molecule was misleading. HCHO is more accurate, showing that formaldehyde is the simplest member of a group of highly reactive compounds, called aldehydes, all of which contain the CHO structure. Many of the other, less famous aldehydes—acetaldehyde, acrolein, crotonaldehyde, chloroacetaldehyde, furfural—are also of environmental concern because of their vast industrial usage. Cigarette smoke, moreover, contains at least ten of them, which will send a roomful of smokers off the health safety charts.

Exposure to formaldehyde at high enough levels causes irritation of the eyes (50–2,000 parts per billion) and upper respiratory tract (100–25,000 ppb) that can become impossible to stand, leading to nausea, vomiting, and dizziness. Numerous studies suggest that formaldehyde triggers asthma. The odor threshold of 50 to 1,500 ppb has nothing to do with a safe level of exposure since most people become somewhat acclimatized to the smell. In 1946, the American Conference of Governmental Industrial Hygienists (ACGIH) recommended an eight-hour average of 10,000 ppb as the limit for formaldehyde in workplace air. Over the years, this has been lowered to 3,000 ppb. The National Institute for Occupational Safety and

Health recommends 1,000 ppb. Several western European nations observe an indoor standard of 100 ppb. During atmospheric inversion conditions over major cities, concentrations in ambient air can reach 50 to 86 ppb, with Los Angeles having suffered peaks of 90 to 150 ppb. Even rural environments register around 0.8 ppb.

In 1979, a sea change occurred in public health policies about formaldehyde when lab rats were found to contract nasal cancer after breathing concentrations of 14,300 ppb and 5,600 ppb in air for six months. Epidemiological studies have turned up limited evidence of similar effects in humans, though some have uncovered higher than normal risk for cancer in other organs. Formaldehyde is therefore classified as a suspected human carcinogen. For this reason, the Consumer Product Safety Commission voted in 1982 to ban urea-formaldehyde foam insulation in houses, where concentrations were averaging 120 ppb. Exposures as high as 4,200 ppb were recorded in mobile homes, in which such insulation was especially common. Although industry lawyers successfully challenged the ban in court, publicity about indoor pollution led to changes in construction materials and techniques. Still, concentrations of nearly 200 ppb have been detected in new public buildings. Most homes contain measurable levels; typical furniture, cabinetry, and only one paneled room will generate 20 to 70 ppb. Houses with particleboard subflooring will measure 60 to 300 ppb two to five years after installation. Concentrations rise on hot humid days.

Formaldehyde concentrations in a room where several people are smoking can reach 250 ppb, with a typical puff of cigarette smoke containing 40,000 ppb. By comparison, the exhaust of a car cruising at forty miles per hour registers about 1,000 ppb (4,000 ppb if it's a diesel engine). Most people cannot tolerate prolonged levels of 4,000 to 5,000 ppb.

One of the neat things about formaldehyde is that if you kill yourself by smoking it, you can also get pickled with it. The chemical's excellent preservative and hardening action on tissue was noted from the start, spawning a variety of patented embalming fluids laced with additional agents—salts, alcohols, lanolin, colorings—to create lifelike cosmetic effects. A proposal by Russian president Boris Yeltsin to bury Lenin's body, which has been on display in Red Square since 1924, brought a protest from the Russian Academy of Medical Sciences that burial would ruin "a unique experiment in preserving human tissues and cells in a constant state for decades."

Formaldehyde has long been used to keep myriad capitalistic products—waxes and polishes, ferns and flowers, hides and textiles, fats and oils, but *not* food—from deterioration by BACTERIA, molds, and FUNGI. So how could we leave this subject without a nod to notorious thief John Dillinger's mythic member, sought by generations of field trippers wandering through the Smithsonian Institution, which along with Albert Einstein's brain and Che Guevara's hands has helped give "preserved in formaldehyde" its deliciously macabre ring.

• freon •

The trademark name for a family of so-called chlorofluorocarbon compounds, or CFCs, introduced by the E. I. DuPont company in the late 1920s and 1930s. They are harmless to people except at extremely high concentrations, but destructive to the Earth's OZONE layer and therefore banned by international agreement.

Freon was invented in 1928 by Thomas Midgley (1889–1944), a Cornell-educated engineer who had developed antiknock leaded gasoline earlier in his career. Searching for a non-toxic, nonflammable refrigerant for the burgeoning air conditioner and home refrigerator market, he discovered that by substituting chlorine or fluorine atoms for hydrogen in ordinary hydrocarbons such as METHANE and ethane, all the desirable features—inert, odorless, minuscule viscosity, boiling point far below room temperature—were cheaply obtained. Hundreds of such compounds were created at DuPont, with those based on methane numbered from 1 to 100 and those from ethane between 100 and 200. Hence "R-12," chemically notated as CF_2Cl_2, the common Freon refrigerant for automobile air conditioners that has become a hot black market commodity.

CFCs were also perfect for use as industrial solvents, degreasers, and especially aerosol propellants. In 1974, Sherwood Rowland and Mario Molina of the University of California published an article in the journal *Nature* showing how chemically stable CFCs, particularly the propellant R-11 ($CFCl_3$), could slowly rise into the stratosphere, be broken down by ULTRAVIOLET radiation from the sun, and thereby release chlorine atoms that interfere with natural production of ozone. In fact, one pound of Freon could destroy 70,000 pounds of ozone. The process remained theoretical until "holes" of abnormally low ozone concentration were discovered over Antarctica in the mid-1980s. From zero at the beginning of this century, the concentration of CFCs in the atmosphere has risen to more than 400 parts per trillion, and the concomitant amount of chlorine in the stratosphere has tripled.

Though present at much lower concentrations than CARBON DIOXIDE and methane, CFCs also contribute significantly to the greenhouse gas alteration of the Earth's heat balance (CO_2, 56%; CH_4, 15%; CFCs, 24%). A molecule of R-12, for example, can trap 15,800 times more heat than CO_2.

In 1987, more than 140 nations signed a treaty, known as the Montreal Protocol, that made the manufacture of CFCs illegal in the United States (except for medication inhalers) and many other countries after January 1, 1996. Importation of foreign-made CFCs was also banned in the United States. By 1993, most new refrigerators and home or car air conditioners were no longer using Freon, but the existence of older units, especially in some 80 million autos, created a bootlegger's paradise. Recharging a car's air conditioner from the shrinking supply of legal R-12 costs upward of $200, ten times the price of a decade ago. The entrepreneurial result: a black market for some 60 million pounds of contraband Freon valued at $1.5 billion, making CFCs second only to smuggled drugs. In 1997, a Florida company was fined $37 million for smuggling four thousand tons of R-12 in three years.

Where does it all come from? The Montreal Protocol allows developing countries to make CFCs until 2010. Consumption in the Third World is expected to double every seven years and soon reach levels attained by industrial nations, according to a United Nations report. Mexico is thus a prime source for illegal Freon in the *Colosso del Norte*. And American companies such as DuPont still turn out CFCs at their plants in India, Brazil, the Netherlands, and elsewhere offshore. The Russian mob—supplied by Russian or former Eastern Bloc factories defying the treaty—runs some of the most lucrative action, tracked by a federal interdiction program called Operation Cool Breeze, but there is plenty of mom-and-pop traffic across the Rio Grande. At the end of the pipeline, service stations cop a hot thirty-pound tank of R-12 for $400, then do recharge jobs at $80 a pound. Cool.

Researchers in Belgium have found that two Freon substitutes, HCFC-123 and 124 (technically known as partially halogenated hydrochlorofluorocarbons), whose use as coolants

and cleaning agents is on the rise because they are less destructive, cause liver damage at high exposures. Factory workers
there developed acute hepatitis after four months in contact
with leaking air conditioner pipes. According to the U.S. Environmental Protection Agency, these HCFCs are actually less
toxic than the CFCs they replace and are necessary while
ozone-gobbling Freon is phased out around the world. They,
too, are scheduled for gradual phase-out, but the Clinton
administration has acted on behalf of manufacturers in resisting
demands from European governments to accelerate the ban.

• fungi •

Merely "mushroom" in Latin, but in taxonomy a group also
including some 50,000 microbic molds, mildews, rusts, and
yeasts that may or may not be unicellular. Without chlorophyll,
they cannot rely on photosynthesis for sustenance, so they need
to grow on what they eat, whether it's the sulfur in a concrete
wall or the decaying organic matter of a tree stump, a mucous
membrane, or a baby's funky bottom.

When a fungal SPORE lands on a moist surface, it cracks open
like an egg so that myriad hyphae, or feeding tubes, can plug
into the surroundings to absorb food. The creature cleans itself
by venting an aerosol of extraneous or excess material, which
can build up to the familiar "musty" smell of a summer cot-
tage. Fungi growing on layers of old oil-based paints binded
with arsenic have actually poisoned people as they released the
unwanted element into stuffy rooms.

Fungi that may cause health problems when they directly col-
onize us are usually opportunistic cowards that need weakened
hosts. They lead benign existences until something—another

infection, chemo- or radiation therapy, diabetes, leukemia, or just the vagaries of biology—hobbles our immune system enough to let them really get going. Others may be confined to certain environments or geographic areas that we encroach upon, such as the *Coccidioides immitis* in dusty regions of the American Southwest that causes valley fever when its spores are inhaled. After the 1994 Northridge, California, earthquake, 203 people contracted the disease, which produces flulike symptoms at worst in otherwise healthy individuals, but deadly pneumonia and meningitis in the immunocompromised.

Aspergillus is as common in the human environment as bread, which it will gladly turn green if you leave a slice out on the table long enough. It loves compost heaps, the insulation in walls and ceilings, household surfaces, and the symbiosis heaven of our mouths and skin. But about 5 percent of kidney and bone-marrow transplant patients come down with bad systemic infections that are hard to diagnose because the fungus is difficult to culture from blood samples. In 1996, *Aspergillus fumigatus,* which can cause respiratory problems, nausea, dizziness, and irritation of the eyes, nose, and throat, was blamed along with airborne spores of *Stachybotrys chartarum* for making the ten-story Department of Transportation headquarters in Washington, D.C., a "sick building."

Blastomyces dermatitidis will infect human lungs and skin, but is a bigger problem for dogs (and may accompany their bites). Most cases occur in the southeastern United States. *Candida albicans,* a yeast found naturally in our digestive tracts and various mucous membranes, can mushroom when faced with a weak defensive system into the disease called thrush, which is common among AIDS patients. But it's also one of the culprits behind diaper rash (in concert with BACTERIA that make irritating AMMONIA from urea in urine) and yeast infections of the female genital tract. A nutritious cousin, *Candida utilis,* was

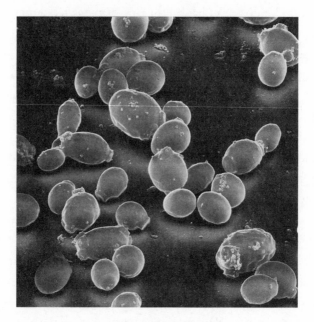

Candida albicans *fungus.*

called upon during the food shortages of World War II and is enjoyed today by truly die-hard Brits as "Marmite."

Cryptococcus neoformans (not related to the protozoan *Cryptosporidium*) is ubiquitous, though it has a special yen for pigeon guano. Usually harmless, when inhaled by an individual with Hodgkin's disease or AIDS it tends to spread into the nervous system and cause deadly meningitis. *Histoplasma capsulatum* is also found almost everywhere, preferring river valley soil that, when dried out, produces dust laced with its spores. The folkname "caver's ague" refers to its predilection for bat-infested caves, where spelunkers run higher than normal risk of infection. Hundreds of thousands of cases occur in the United States every year—particularly in Arkansas, Tennessee, Missouri, and Kentucky—bringing problems ranging from flulike symptoms to pneumonia. For AIDS patients, however, it is deadly.

For you avid gardeners, *Sporothrix schenckii* dwells on rose-bushes, barberry, and in peat moss or other mulches. It can cause skin ulcerations and, if inhaled, pneumonia.

Then there is ringworm, or tinea, which is actually not due to a worm but a fungus named *Trichophyton*. Various species prefer the body, the athlete's foot, the nails, the scalp, the crotch, the beard—basically whatever is dark, moist, and grubby.

The fungal infections that most worry doctors today are the ones that spring forth because of aggressive medical procedures like chemotherapy and organ transplants, which undoubtedly prolong life but hobble natural defenses against invading microbes. Between 1980 and 1990, nosocomial (acquired while in hospital) fungal infections doubled. Getting rid of them is not easy. They are much closer to our own cellular makeup than bacteria, so few drugs are able to squelch them without regrettable side effects.

Yet we have the fungus family to thank for lifesaving penicillin and other antibiotics, delectably veined cheeses, brewer's yeast (*Saccharomyces cerevisiae,* olé!), and succulent morels. All in all, we should be glad we are able to live among them.

• gamma rays •

A type of radiation that is analogous to X RAYS but more powerful. Both are highly energetic forms of light, with much shorter wavelengths than the visible spectrum. The French physicist Paul Villard discovered gamma rays in 1900, choosing the next letter of the Greek alphabet after the already-taken ALPHA RAYS and BETA RAYS.

Gamma rays are spawned by nuclear interactions in violent cosmic phenomena, such as supernovas and black holes, but the

Earth's atmosphere shields us from such distant sources. A gamma-ray halo, trillions of miles thick, surrounds the Milky Way galaxy. Its photons are a billion times more energetic than those of visible light. In February 1997, the Dutch-Italian satellite Beppo-Sax was able to pinpoint for the first time the origin of a tremendous gamma-ray burst (GRB) beyond our galaxy at least 7 billion light-years from Earth. The existence of GRBs has been known since the late 1960s, when they were noticed accidentally by Vela satellites designed to detect clandestine nuclear weapons explosions in space, which would also release intense gamma rays. For seconds or minutes, they flash in the sky about three times a day with more energy than the Sun will release in its 10-billion-year life span. The gamma-ray pulse that struck Beppo-Sax lasted about eighty seconds from a position halfway between Alpha Tauri and Gamma Orionis in the Orion constellation. Possibly caused by stellar material orbiting a black hole or the collision of two superdense "neutron" stars (collapsed cores of supernovas, only a few miles in diameter), the GRB's light must have traveled for about half the age of the universe before it bathed the satellite's sensors.

On Earth decaying radioactive elements also release gamma rays, which are far more penetrating than alpha or beta radiation. Alpha and beta particles of a given energy have a definite range and will be completely absorbed by a certain thickness of material. But some gamma rays will go through even very thick shielding. Atomic weapons researchers created the notion of *tenth-value thickness* as a measure of how well various substances block gamma rays: one tenth-value thickness decreases the radiation hitting the material by a factor of ten. Thus, steel has a tenth-value thickness of 3.3 inches; concrete, 11 inches; earth, 16 inches; water, 24 inches; and wood, 38 inches. Perhaps it was this concept that emboldened T. K. Jones, an undersecretary of defense for nuclear forces in the Reagan administration, to proclaim in 1981 that a nation could survive nuclear war "if there are enough shovels to go around."

Compared with alpha and beta rays, gamma rays are less damaging to living tissue, primarily because they have no substance and zoom through at the speed of light. But they are still quite capable of biochemical mayhem as they ionize—that is, strip electrons away from—neutral atoms or molecules. In the Hiroshima and Nagasaki bombings, gamma rays are believed to have been the principal radiation threat. An acute, whole-body gamma-ray dose of about 100 rads would probably kill half of the people exposed within a month, without medical treatment. The National Academy of Science's BEIR-V report of 1990 estimated that "if 100,000 persons of all ages received a whole-body dose of [10 rads] of gamma radiation in a single brief exposure, about 800 extra cancer deaths would be expected to occur during their remaining lifetimes in addition to the nearly 20,000 cancer deaths that would occur in the absence of radiation." Nearly all natural gamma emitters of biological concern also give off alpha and beta radiation, in which case the gamma ray does not represent a large share of the total dose.

A rather macabre term from weapons research for rating the destructiveness of different types of radiation is *relative biological effectiveness,* or RBE. By definition, the RBE for X rays and gamma rays is approximately unity; for the same dose of beta rays, roughly 10; and for alpha rays, 20.

• God •

He was a wise man who invented God, wrote Plato in *Sisyphus.* Polls find that many professional scientists continue to believe in a deity, though not necessarily one with all the traditional biblical trappings. Their faith is nonetheless higher than William Blake's brand of pantheism:

> *The pride of the peacock is the glory of God.*
> *The lust of the goat is the bounty of God.*
> *The wrath of the lion is the wisdom of God.*
> *The nakedness of woman is the work of God.*

It goes beyond the simple respect expressed by the Polish proverb: God can shave without soap. And it is without cynicism, unlike Pope Xystus I, who declared that "God is not the name of God, but an opinion about Him."

Why would any educated person believe in God this far into the scientific age, parental guilt trips aside? Perhaps God is sometimes an artifact of scientific inquiry itself—the further one digs, the more of nature encompassed by theory, the grander the unification, the less likely it seems that everything could be happenstance. Even Immanuel Kant, the great demolisher of intellectual proofs of God, rather grudgingly admitted that "God's perfections are marvellous (but not lovable)." He respected the argument that order in the universe suggests purpose, but would consider only the possibility of an Architect, not a full-blown Creator. It was Kant, of course, who believed that all planets were inhabited, and that the most distant ones had the best inhabitants.

We can be thankful that God will always be invisible, a state that guarantees perpetual belief among mortals. The Welsh say that there are three things that only God knows: the beginning of things, the cause of things, and the end of things. They seem like natural-born physicists. And the Irish, who counsel that God is not as severe as He is said to be, must be engineers. For these hubristic musings, God will forgive me, because, as Heinrich Heine gasped on his deathbed, "It is His trade."

• **gravity** •

"Can you bind the chains of the Pleiades, or loose the cords of Orion?" Nope, sure can't. Thus did GOD put Job in his place, because gravity is the one force in nature that cannot be switched off, shielded, neutralized, reversed, released, or otherwise messed with, the Wonderbra notwithstanding. It is omnipotent and omnipresent, just like You-Know-Who.

Those anonymous Hebrew mystics who produced the Book of Job in the sixth or fifth century B.C. knew no more about gravity than a poor babe who falls out of its manger, of course. Around 330 B.C., Aristotle surmised that the four so-called elements—earth, air, fire, and water—tended to settle at increasing speed into their natural resting places. Not much better than the mystics. So let us leap ahead to Mikolaj Kopernik, a.k.a. Nicolaus Copernicus, the Polish physician and astronomer who espoused a heliocentric solar system, in his *De revolutionibus orbium coelestium* of 1543: "Gravity is nothing else than a natural force implanted by the Creator of the world into its parts, so that, coming together in the shape of a sphere, they might form a unified whole." Or Giordano Bruno, the radical humanist who believed the universe was infinite and composed of many solar systems (and who was therefore burned at the stake for pantheism in Rome's Campo de Fiori), in his *Del infinito, universo, e mondi* of 1584: "Gravity and lightness are only attraction and flight—nothing is naturally heavy or light." Or Johannes Kepler, who charted the elliptic planetary orbits, in his *Astronomia nova* of 1609: "Gravity is a natural corporeal attraction between bodies toward a connection, so that the earth attracts a stone much more than a stone attracts the earth." Lo and behold, these great Renaissance scholars were still at the level of tribal poets when it came to thinking about gravity.

Facing the inevitable at about 32 feet/sec².

Though Isaac Newton's *Philosophiae naturalis principia mathematica,* written between 1684 and 1687, and Albert Einstein's general theory of relativity, published 230 years later, explained how gravity acts, we remain in the desert on why this force works. But never mind; its action alone is hard enough to grasp.

Newton and the falling apple is rightfully the most famous apocryphal tableau in all of science. His influence on modern society is—forgive us, John—on a par with Jesus'. Mastering the methodology of Newton's law of gravitation is the starting point for any career in science or engineering. He reasoned that gravity acts on material objects "at-a-distance" throughout the universe without requiring contact or an intervening medium and found stunningly simple equations for describing its effect. If the heavens were thus subject to the same natural dictums as the Earth, then perhaps social principles such as equality should be universal, too. Or so European philosophers such as Locke and Voltaire began to think as Newton's work resonated beyond scientific circles. As for the heavy question *why do things attract each other?*—he was quite content to believe that only God knows, thus leaving one foot safely in the Bible.

Newton's law says that the force of gravity between two objects is equal to the product of their masses divided by the square of the distance between them, all multiplied by a numerical constant. In symbols this is written $F = G(M_A \times M_B/R^2)$, where G, a fundamental quantity of nature, is defined as the magnitude of attractive force between two one-kilogram masses separated by one meter. G was first measured in 1798 by the English physicist Henry Cavendish (1731–1810), whose family later endowed the famous physics laboratory at Cambridge University. The important thing to note about G, which equals 6.67259×10^{-11} in metric units, is how small it is. The force of gravity is minuscule unless at least one of the masses is gigantic.

For example, it takes the entire bulk of the Earth (about 10^{24} kilograms) to accelerate an object downward near the planet's surface at the same rate that a sprinter achieves with muscle power at the start of a race. Whenever you pick something up, you conquer this tremendous mass. Also, gravitational force gets quickly smaller as the separation distance increases, though it never reaches zero. Of the four basic forces now known to science—electromagnetic force, strong nuclear force, weak nuclear force, and gravity—gravity is by far the most feeble, yet it binds the Pleiades.

To perceive his laws of motion, Newton postulated the existence of "absolute space," a fixed frame of reference that is not moving relative to anything else. He understood the artificiality of this assumption, but almost two and a half centuries passed before someone came along who was so bothered by it that he invented a whole new approach. Einstein did away entirely with the Newtonian notions of force and action-at-a-distance, reverting to a Grecian geometric point of view and positing instead that all objects travel through a four-dimensional realm defined by the three dimensions of normal space, plus time. To get from one point to another, they simply follow the shortest path. Masses make the streamlines of spacetime curve inward—barely at all at great distances, but sharply at short range. Why? Einstein didn't say.

For most situations in human experience, the methodologies of Newton and Einstein give results that differ only trivially. When NASA sends astronauts to the Moon or robotic spacecraft to Mars, Newton is in control. To explain Galileo's famous (though unconfirmed) observation at the Tower of Pisa that two different masses fall at the same rate, Newton talked about the heavier object's greater inertia, whereas Einstein chose to say that the flow of space-time carries each mass along together like flotsam in a river. They each covered Galileo.

On the scale of cosmic events, however, where gravity is titanic and matter is moving close to the speed of light—such as during the collapse of stars into black holes or in the first few moments after the Big Bang—Einstein's approach is revealed as more fundamental. The first confirmation of his insight involved explaining a small discrepancy between the orbit of Mercury as predicted by Newton's equations and the observed path. (In 1898, the American astronomer Simon Newcomb had calculated the difference to be *exceedingly* small—forty-three seconds of arc per century, or about the angle subtended by a mouse one kilometer away.) Without the general theory of relativity, astronomers had been driven to assume there must be an undiscovered planet closer than Mercury to the Sun, which they called "Vulcan," causing this perturbation. But Einstein's way of dealing with gravity relegated Vulcan to a future sci-fi television hit. Then came Sir Arthur Stanley Eddington's dramatic demonstration of "gravitational lensing" during a solar eclipse in 1919, whereby the mass of the Sun distorts starlight precisely as Einstein predicted.

One crucial feature of Einstein's work has yet to be confirmed: gravity waves. Whenever great masses change motion—say, when stars collide—they should generate waves of energy that actually stretch and shrink the three-dimensional geometry of space. This effect would be infinitesimally small, however, and none of the detectors built so far on Earth have found any trace of it. The European Space Agency has proposed orbiting three satellites around the Sun linked by laser beams, so that gravity waves passing across the array would squeeze one side of the resultant triangle while stretching another enough to be measured. The 5-million-kilometer "legs" of the triangle would be distorted by less than the diameter of an atom.

Indirect evidence of gravity waves is thought to be already in hand from observing the energy lost since 1974 by a certain binary pulsar, a pair of neutron stars spinning around each other on their way toward ultimate collapse some 400 million years from now. Only the escape of energy via gravity waves could account for the magnitude of loss so far, as calculated from the shortening separation between them.

But no one knows *why*. If gravity is a quantum mechanical force like the other three fundamental forces, there should be a corresponding force-carrying "particle" like the PHOTON. Physicists have a nice name for it—the graviton—but no evidence of its existence.

As for black holes, perhaps the greatest gift of theorists to twentieth-century science fiction writers, a half dozen or so candidates have been identified since the concept gained credence in the late 1960s. They are presumed to be formed when old stars cool and collapse upon themselves until all that remains is a pinpoint of gravity so strong that neither light nor any other kind of radiation, let alone mass, can escape. The first one discovered, in 1971, is part of a powerful X-ray emitter called Cygnus X-1, where a visible star is apparently slipping gradually down the maw of the invisible black hole. Our Milky Way galaxy itself is thought to have a colossal black hole at its center. (We're too distant to get sucked in, unfortunately.)

Although the term *black hole* was coined in 1967, the possibility of an object so massive and compact that light could not escape from it was first imagined in 1783 by the Reverend John Mitchell, who devised Henry Cavendish's method for determining the value of G. If the Sun could be compressed to a diameter of six kilometers, it would become such a thing. During World War I, the German physicist Karl Schwarzschild calculated—while dying in a military hospital—what would happen if a gravitational field were strong enough to make space curve in on itself and be cut off from the rest of the universe.

The Hubble Space Telescope has provided the most direct evidence for black holes, especially since the installation of its imaging spectrograph. By measuring the velocity of gaseous streams within twenty-six light-years of the center of galaxy M84 (about 50 million light-years from Earth) to be as high as 880,000 miles per hour, Hubble scientists confirmed a black hole with mass at least 300 million times greater than the Sun's.

Though Orion's cords are as tight as ever, astronomers and physicists remain vexed by the fact that there is not enough visible mass in the universe to account for the gravitational pull required to slow the universe's expansion since the Big Bang of creation. About 90 percent is missing, actually. This "dark matter," as they call it, could be stars too dim to be seen (known as massive compact halo objects, or machos). Or it could be NEUTRINOS, particles of infinitesimal mass that rarely interact with matter. Or it could be something that so far exists only in scientists' imaginations—stuff with unknown properties aptly dubbed wimps, for weakly interacting massive particles. Wimp detectors are starting to appear at various sites around the world, but the task of their builders is rather daunting, since they don't know what they're looking for. Roll over, Lewis Carroll.

Let us end on solid ground. Consider that since the Earth is not a perfect sphere, gravity is stronger at the poles than at the equator. Why then aren't skinny girls in bikinis on tropical beaches flung off the spinning globe by centrifugal force? Because gravity is still 300 times stronger, praise God. There is no record that Newton considered this problem, but he did wonder what gravity would feel like during a journey to the center of the Earth. In the *Principia,* he showed that the pull would decrease continuously, since the only mass that matters is what remains below you. At the center, you'd be weightless. (Actually, because the planet is pear-shaped, there would be a minuscule tug toward the south polar region). A frictionless slide to the opposite side of the Earth and back again would take ninety minutes, with or without a bikini.

• hydrogen sulfide •

An invisible gas that smells like rotten eggs and is as poisonous as cyanide. Its darksome toxicological history traces back to the sewers of Paris, as does its literary reputation. There is no more vivid description of H_2S's effect on humans than in Victor Hugo's *Les Misérables* (1862): "Slow asphyxia by uncleanliness, a sarcophagus where asphyxia opens its claws in the filth and clutches you by the throat; fetidness mingled with the death-rattle, mud instead of sand, sulphuretted hydrogen in lieu of the hurricane, odure instead of the ocean!" Calling the sewer system the "Intestine of the Leviathan," Hugo reflected the morbid obsession of Parisians with *le tout-à-l'égout.*

As a natural product of putrefying organic matter, hydrogen sulfide was well known to ancient philosophers, who were apparently so impressed with its noxious power that they named it "divine water" from the Greek word *theion,* meaning "divine" or "sulfurous." The Christian relegation of this particular stench to the Devil was yet to come. H_2S was first analyzed in the 1770s, particularly by the Swedish apothecary and chemist Carl Wilhelm Scheele (1742–1786), who referred to it as *Schwefelluft* ("sulfur air") when he was feeling scientific or just *stinkende* ("stinky" or "fetid") after a long day in the lab. Though he had also discovered hydrogen cyanide and described its odor and taste, he did not comprehend how toxic these gases were and was lucky to live past forty. He died from the gout.

During the time of Scheele's work, sewer gas leaks caused fatalities regularly in Paris, and the effects of this as yet unidentified scourge on the eyes of privy cleaners had been known empirically for decades. By the end of the eighteenth century, various researchers had proved the presence of hydrogen sulfide, but lethal accidents continued for decades in Paris and London. Sulfur miners, including children, were also prone to H_2S asphyxiation. The first reported American death occurred in 1851—not in a city sewer or mine but in an outhouse. Despite the growing body of scientific evidence about the gas's toxicity to animals, plants, and people, bizarre clinical uses endured, such as injecting it into the rectum to treat pulmonary diseases. What comes out should be put back in (see FLATUS).

A 1929 account of hydrogen sulfide pollution in the Texas oil fields noted that fifteen to thirty deaths had occurred in two years, all wildlife had disappeared, and silver coins turned black with tarnish. In 1950, 320 people were hospitalized and 22 died in Poza Rica, Mexico, after a refinery flare malfunctioned and released unburned hydrogen sulfide across a community of bamboo huts. Until the advent of air-quality regulations, the American emergency medical record of this century was filled with reports of acute H_2S poisoning among sewer, refinery, and farm workers. The reek of the local sewage plant for miles downwind was a familiar fact of life in large cities. In the 1960s and 1970s, biodegradation of industrial waste lagoons was still known to generate high enough ambient concentrations (300 parts per billion) to trigger mass episodes of nausea, breathlessness, and headaches in Terre Haute, Indiana, and Alton, Illinois.

Today H_2S remains a by-product and occupational hazard of many industrial processes, including oil and gas drilling, petroleum refining, paper manufacturing, fish meal production, rayon production, leather tanning, the preparation of various sulfur compounds, and the production of heavy water as a moderator for nuclear reactors. In general, thanks to environmental laws, the gas is no longer an everyday pollutant of urban air. Its background level in the atmosphere is about 0.05 ppb, well below the odor detection threshold of about 100 ppb, but enough to discolor house paints pigmented with white lead. The smell becomes clearly offensive at around 3,000 ppb, will cause serious eye injuries at 50,000 ppb, olfactory paralysis at 150,000 ppb, and imminent death above 300,000 ppb. The federal limit is 10,000 ppb for an eight-hour workday—not a job environment that anyone should aspire to.

• **infrared** •

The region of the electromagnetic spectrum between visible light and MICROWAVES. It begins just below what we can see as the color red—from about three-quarters of a micron in wavelength to one millimeter—hence the Latin prefix *infra*. William Herschel (1738–1822), an English astronomer and musician of German birth who discovered Uranus, first detected infrared radiation when he showed in 1800 that most of the heat in a prism's beam comes from beyond the red band. Despite this knowledge, he believed that the Sun was an orb much like Earth, surrounded by luminous atmosphere, and inhabited.

Infrared rays are usually associated with heat, because an increase in temperature is easy to detect when molecules absorb IR and start to vibrate. Conversely, all objects with temperatures above absolute zero emit IR. The higher the temperature, the shorter the wavelength. Thomas Edison's first bulbs were hot but rather dim because the filaments he used—carbonized strips of bamboo—operated at temperatures where far more energy was radiated as infrared than as visible light. Even modern incandescent bulbs convert electricity to light with only 5 percent efficiency. If we could see infrared rays naturally, there would be no such thing as a dark night, because the Earth's own heat would illuminate everything like a full Moon.

Some 55 percent of the Sun's energy is in this thermal region of the spectrum. Of the 340 watts per square meter of solar energy that reaches the Earth's outer atmosphere, about 100 watts are reflected back into space, 80 watts are absorbed by the atmosphere, and 160 watts are soaked up by the oceans and land masses. In turn, the planet's surface gives off heat by infrared radiation at 390 watts per square meter, 320 watts of which gets reflected back down by clouds and "greenhouse" gases (mainly WATER VAPOR, CARBON DIOXIDE, and METHANE), which thus provide two times more heating of the surface than comes directly from the Sun.

Generally, any biological damage from infrared radiation is due to excessive heat, running the gamut from obvious burns to subtle effects such as temporary oligospermia (deficiency of sperm cells) from tight underwear. A healthy adult typically radiates about 500 watts of IR, triggering "motion detectors" and illuminating the "sniperscopes" of World War II. Rattlesnakes are exquisitely sensitive to infrared, with two facial organs that can detect temperature differences of a few thousandths of a degree sending nerve impulses to opposite sides of the brain, where a stereo heat image is produced for precise nighttime strikes. How much would the generals pay for a targeting system like that?

• **krypton** •

The ninth most common gas in unpolluted air, with a concentration of about one part per million by volume. In Greek, *kryptos* means hidden, an image chosen by the British chemist Sir William Ramsay (1852–1916) in 1898 when he discovered this rare element during a quest for missing noble (or inert) gases in the periodic table. That same year, he also found neon (from Greek *neos,* new) and xenon (from *xenos,* strange). In 1900 he filled in the last box with a phosphorescent gas he called niton, from Latin *nitere,* to shine, but the name was later dropped in favor of RADON. Nonetheless, he won a Nobel Prize for his work in 1904.

Like the far more abundant argon and neon, krypton is sometimes used in fluorescent lights, but the difficulty of separating it from air makes it generally too expensive for consumer products. It solidifies at −314° F—hardly worth the effort, since a chunk of it will not rob Superman of his powers. Perhaps its most enduring role is as the international standard for defining the meter: 1,650,763.73 wavelengths of the orange-red spectral line of one of its six stable isotopes, krypton-86. At least this seems preferable to the king's arm.

• laughing gas •

Nitrous oxide, N_2O, a colorless gas with a sweet taste and smell that will give you the sillies before knocking you out. It was a common anesthetic for light surgery and dentistry until the 1970s, when studies found that female dental assistants were suffering twice as many miscarriages as unexposed women. A 52 percent increase was also observed among dentists' *wives.* Not funny.

N_2O was discovered in 1772 by Joseph Priestley, the English chemist, Congregationalist minister, and possibly repressed party animal who also invented carbonated beverages (see CARBON DIOXIDE). Its ability to induce euphoria quickly led to the popular nickname as "revels" were organized for amusement. Another chemist, Humphry Davy (1778–1829), noticed that laughing gas also brought reduced sensitivity to pain. He

Laughing gas, a prescription for scolding wives, 1830.

proposed to no wide avail that it be inhaled during surgery. (Davy soared to the top of his academic profession, was knighted, and retired at the age of thirty-four. He married a rich widow and spent the rest of his life "traveling on the continent," as people used to say. The role laughing gas played in his lighthearted approach to careerism is unclear.) Nitrous oxide was not used as a dental anesthetic until 1844, when an enterprising Connecticut tooth drawer named Horace Wells (1815–1848) witnessed a public exhibition of its jolly effects and then tried it while having one of his own yanked. After offering laughing gas to several clients, he arranged a demonstration for a medical class at Harvard, where he was booed out of the hall as the patient screamed in agony. The modern era of painless surgery, or "anesthesia" as Oliver Wendell Holmes christened it, thus had to wait a few more years until the acceptance of ether and chloroform.

Properly classified as an asphyxiant, laughing gas's lingering good-time reputation was finally ruined by reports that the half-and-half mixture of nitrous oxide and oxygen commonly administered for continuous sedation causes temporary bone marrow loss. A few deaths from anemia set off loud alarms. But it was the evidence of reproductive hazards that changed N_2O from entertainment to abuse. Children of female dental assistants showed a 50 percent higher incidence of birth defects than those of unexposed assistants. A 25 percent increase was found in a survey of the offspring of 49,585 *male* anesthesiologists, compared with children of male doctors who did not work in operating rooms. Nitrous oxide's teratogenic effects were traced to its destruction of vitamin B_{12}, which is needed for cell division and DNA production.

Although reaction-time performance tests suggest that laughing gas doesn't really kick in until it reaches concentrations in the air between 80,000 and 120,000 parts per million, a few studies have noticed mood changes at 50 ppm, which is the federal limit. People really did show up at parties with bottles of the stuff, probably obtained from ice-cream parlors where it was used as a foaming agent for whipped cream. Toxicology handbooks describe the effects of nitrous oxide on the central nervous system in terms of paresthesias, impairment of equilibrium, numbness, and difficulty in concentration. In other words, you start to feel prickly all over, stumble around, and go gaga. This is what parties are for, after all, and the impotency that also accompanies high exposure was seldom faced by revelers in the pre–sexual revolution heyday of laughing gas.

Nitrous oxide is ubiquitous in the air because of natural bacterial action on nitrates in soil. The steep rise in use of nitrogen fertilizers since World War II has pushed more and more into the Earth's atmosphere, where it lingers for 150 years. Its preindustrial concentration was about 288 parts per billion; today it measures around 310 ppb. Because it is also generated artificially by the burning of petroleum products, its concentration is increasing by about 0.25 percent a year. In the troposphere it acts as a relatively minor greenhouse gas, but higher up in the stratosphere it contributes to the production of nitrogen oxides that figure heavily in the destruction of OZONE.

• **methane** •

A well-fed dairy cow jets about 500 liters of this gas every day—out through its mouth as a BURP, not from the other end. Count Alessandro Volta, an Italian physicist after whom the unit of electrical potential was named, is credited with first isolating methane in 1776, but he was not known to consort with livestock.

There are about 4.5 trillion (4.5×10^{12}) kilograms of methane in the Earth's atmosphere, with slightly higher concentrations in the Northern Hemisphere than Southern because of agriculture. About 5 percent of the worldwide energy from photosynthesis goes to generate methane. Its presence as the most abundant organic gas has been known since the 1940s, when the physicist M. J. Migeotte found that it was absorbing light in the INFRARED region of the solar spectrum. Besides animal flatulence, the primary sources of methane are termites, which also produce it in their guts, and so-called marsh gas released through microbiological rotting of organic matter in swamps and rice paddies. Poking a stick into the muddy bottom of a pond that gets seasonal deposits of leaves and other vegetation will release bubbles of methane formed by BACTERIA known as methanogens, which also inhabit those cattle guts. "Will o' the wisp," a flame of burning methane dancing across a peat bog, is one of the rare elements of folklore (Irish, in this case) that is a genuine phenomenon. The same process can be imitated industrially, using farm wastes or sewage, to create the "biogas" used as a power source in Asia and Africa. Significant amounts of methane are also released from landfills, mines, sewage treatment plants, natural gas wells, and pipelines, and around electrical power transmission lines.

For hundreds of years, the amount of methane in the atmosphere has been slowly increasing due to animal husbandry, destruction of forests (which results in a termite population boom), and other human habits. Analysis of air bubbles trapped in deep polar ice shows that methane levels were constant (at about 0.7 parts per million by volume in the troposphere) from at least 25,000 B.C. to the end of the sixteenth century, when concentrations started to ramp upward. Around 1918, the annual increase accelerated and has stood at about 1.7 percent per year since 1965, producing concentrations today of around 1.7 ppm. Because methane is one of the "greenhouse" gases that regulate global temperatures (the others are WATER VAPOR, CARBON DIOXIDE, chlorofluorocarbons, and nitrous oxide), more and more methane has made the world hotter by 0.4° F since the beginning of the industrial revolution. There are no signs of it stopping, so plan your grandchildren's wardrobes accordingly.

A bit of simple math and chemistry drives home how human industry can outstrip something as seemingly limitless as the planet's atmosphere. As methane reacts readily with oxygen, CARBON MONOXIDE is produced at the rate of about 8×10^{14} grams per year. (This is a lucky reaction, since without it the concentration of oxygen in the atmosphere would rise from a safe and steady level of 21 percent to a highly combustible 25 percent in about 48,000 years.) These molecules last about two to three months before converting to carbon dioxide, thus producing about 3.4×10^{14} grams of carbon annually as CO_2. Yet the worldwide release of CO_2 through burning fossil fuels is far higher, at about 53×10^{14} grams of carbon per year.

Methane's status as the main component (about 85%) of natural gas underlines its usefulness as a fuel. The combustion of one methane molecule—which consists of four hydrogen atoms bonded to a single carbon atom—produces 8 electron volts of energy. As a point of comparison, the fission of one uranium-235 atom produces 200 million electron volts. The separation of these two processes by eight orders of magnitude helps explain why a certain stunt performed by flatulent schoolboys has not eradicated the species.

Methane may be said to have contributed to world culture through its historical role in ensuring the fortune of Alfred

Nobel. The inventor of dynamite would not have found a lucrative enough market in time to endow the Nobel Prizes had coal miners not considered his explosive far safer than gunpowder for blasting purposes. Dynamite's quick, cool flame was unlikely to ignite the deadly methane–coal dust mixture, called firedamp, in subterranean tunnels that lacked a canary or two.

The comparative methane production of various animals may be of interest to zoogoers. Camels blow away the competition at 59 kilograms per year, followed by giraffes and buffalo (50 kg), moose (31 kg), elephants (26 kg), and hippos (17 kg). Humans—hard as it may sometimes be to believe—are at the extreme low end of the scale (0.05 kg), just below warthogs (1 kg). Don't worry; methane is totally odorless unless inhaled under high pressure.

• methyl chloride •

A poisonous, colorless gas that possesses what toxicologists describe with uncharacteristic poeticism as an "ethereal" smell. It is an antiknock fuel additive, so we might try to evoke that ever-so-slight, dreamy dizziness from gasoline fumes when filling up the tank on a summer day. Memories are made of CH_3Cl.

After depressing your central nervous system, methyl chloride damages your liver and kidneys. You can die from either quick exposure to high concentrations or prolonged breathing of low levels. The former kind of demise is highlighted by several hours of feeling fine, then progressive drowsiness, staggering, slurred speech, vomiting, diarrhea, fever, convulsion, paralysis, and coma. If you somehow survive all that, neurological problems have been known to arise as long as thirteen years later. The slow-type exit involves confusion, emotional instability, insomnia, blurred vision, vertigo, tremors, and anorexia. If this seems unusual soon enough and you get completely away from the gas, you might recover in three months. Take your pick.

The federal limit is fifty parts per million. During that ethereal fill-up you might get close.

• microwaves •

Between RADIO WAVES and INFRARED light on the spectrum of electromagnetic radiation lies the territory of microwaves. They range from 0.3 GHz to 300 GHz (gigahertz, or billion cycles per second) in frequency, just short of human perception. These are all "non-ionizing" forms of energy, which means that they are not strong enough to cause damage to living tissue by knocking electrons off atoms or molecules the way X RAYS and GAMMA RAYS do. But microwaves can agitate electrons to the point where considerable heat is generated within irradiated material—*violà* microwave ovens. If they can pop popcorn, they can cook your goose, so to speak, hence safety codes for manufacturing myriad devices that emit microwaves: cellular telephones, pagers, VDTs, burglar alarms, radar, et cetera.

The skin, eyes, and testicles have long been known to be especially sensitive to the thermal effects of microwaves, with cataracts a well-established clinical effect. Altered nervous system and cardiovascular functions have been found in some occupational studies. Headaches, insomnia, irritability, and lassitude are associated with chronic overexposure, especially at powers above ten milliwatts (one-thousandth of a watt) per square centimeter of body surface. A study of 4,000 former employees at the American Embassy in Moscow, which was bathed in microwaves several hundred times weaker than 10 mW/cm^2 by the Soviets for decades in an apparent attempt to disrupt communications, found no short-term risks to health.

Cellular phones transmit microwaves at power levels that range as high as 0.6 W. It is therefore possible to exceed the exposure limits to radiofrequency energy recommended by various professional groups—10 mW/cm^2 for one minute, less for longer durations, for example—particularly since the device's antenna is held virtually in contact with the head. Contrary to the faddish habits of many motorists and pedestrians, these popular gadgets are not for long yakkity-yaks.

• mites •

A ubiquitous group of arachnids (a class that also includes spiders, ticks, scorpions, and daddy longlegs) with over 30,000 known species and probably a million-odd more waiting to be identified. This term for tiny crawlies comes down to us through the unwashed civilizations of northern Europe, tracing back through Old English to prehistoric German, *miton,* meaning "something cut into small pieces." The singular also means "little bit," of course, originally used in Dutch to refer to a coin worth one-third of a penny. The latter sense informed the phrase "widow's mite," even though poor old women in medieval Flanders undoubtedly possessed more of the bugs.

Strictly speaking, mites are not invisible to the human eye. We just have to look hard to find them, which may amount to the same thing. The most minuscule are less than one sixty-fourth of an inch across, but there are some big bruisers over an

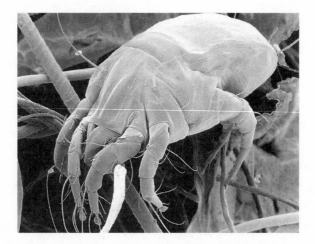

Dust mite, marvel of adaptation.

inch long. As with BACTERIA, most mites are harmless, many are beneficial, a relative few cause problems when their populations burgeon. And like microbes, they occupy a seemingly endless range of habitats, including the oceans and deep underground. Acarologists, as those who study mites call themselves, are richly rewarded intellectually for enduring the *yuch*s of repulsed laypeople.

Eight-legged and sightless, mites live on their own or in parasitic bliss with myriad other creatures. The latter sort may thrive externally or internally on plants, mammals, birds, reptiles, insects, or other invertebrates. They have been found in the lungs, bladder, and intestines of vertebrates, as well as on the skin. Various species have evolved complex appendages for grasping, piercing, and sucking their preferred food. Some carry their young internally. Many mites "drink" via a substance in their leg joints that absorbs humidity. Numerous species exhibit a clever trait known as phoresy, a nonparasitic relationship whereby they hitchhike on unfazed hosts in order to find someplace suitable for the mitey good life. If environmental conditions become too stressful, some even molt into a highly specialized form called a hypopus to undertake such quests. The pilgrim is heavily armored, extra tolerant of dryness, and especially adept at hanging on. Hypopi wait for neighborly beasts on the move, climb aboard, disembark when they reach a fresh habitat, and then resume their normal life cycle. Mites also spawn hypopi on a seasonal basis as conditions become best for reproduction.

A sterling show of phoresy is put on by so-called hummingbird flower mites, which feed and mate on plants pollinated by those peripatetic fliers. They scurry around from one newly opened bud to another, but when they need to find a whole new plant, they hop inside the nasal cavity of a roving hummer. Sometimes more than one species of mite will ride in the same bird's nose, but each gets off only at the kind of flower it prefers, thus avoiding the level of congestion humans endure at JFK or Heathrow.

Workaday collaborations between mites and insects are more common than hummingbird-style airlifts, or at least better known to science. They have been evolving for perhaps 100 million years. (One amber fossil of a mite whose modern descendants consort with beetles dates from the Oligocene period 25 to 40 million years ago.) Bees and other nest makers are especially entwined with providing food and transport to mites, usually without detriment to themselves. The mites steal honey, of course—a phenomenon known by the marvelously psycho term *cleptoparasitism*—but also perform household chores, such as cleaning up dead bees or the FUNGUS on feces and other organic clutter in the nests. They may even munch BACTERIA.

Only a small minority of mites feed directly on their living hosts, an indignity suffered by the highly social bumblebees and honeybees (solitary bees would be a dead end), which counter with careful grooming. Phoresy-stage mites know exactly where to situate themselves to avoid being brushed off, and different species on the same bee maintain separate seating areas. Carpenter bees have even developed a special pouch, called an acarinaria, for carrying one kind of mite that kills another that eats the bee's brood.

Birds and mammals, but not fish, are parasitized by mites worldwide. Only penguins are free from feather mites, for example. Itch mites, *Sarcoptes scabiei,* can infest the skin of almost all domesticated animals (except cats, which are the target of another mite species, *Notoedres cati*) with potentially lethal mange, perhaps due originally to contact with humans. While scabies is spread by *S. scabiei* from person to person usually by sleeping in the same bed, polite texts skirt around this perverted possibility for human-animal transmission by speaking of "husbandry and domestication practices." In people and animals, female itch mites tunnel under the skin to lay their eggs, thus causing intense scratching.

Humans also carry many other species of medical concern. *Liponyssus bacoti,* for example, may spread murine typhus, though the equally difficult-to-spot body louse, *Pediculus humanus* (COOTIE), and rat flea, *Xenopsylla cheopis,* are better-known vectors. Mites of the family Dermanyssidae, which includes the house mouse mite, transmit rickettsial pox, tularemia, and—via birds—a painful skin irritation. *Ophionyssus natricis,* the snake mite, causes fatal diseases in captive snakes but also infests rats and people. A close cousin,

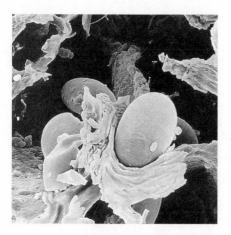

Got the itch. Eggs of scabies mite in skin.

O. bacoti, called the rat mite even though it also likes other small mammals, happily bites people living in rat-infested buildings. Another rodent mite family, Laelapidae, transmits plague. House DUST mites, *Dermatophagoides farinae* and *D. pteronyssinus,* convey powerful ALLERGENS that trigger asthma. Chiggers, which are mite larvae of the family Trombiculidae, also known as redbugs, transmit scrub typhus (tsutsugamushi fever) in the Asiatic-Pacific area and Q fever. *T. alfreddugesi* is the most common chigger in the United States that bites humans, raising small red itch welts. But not all mites that parasitize humans are threatening. *Demodex folliculorum,* for example, can be found in the pores on almost anyone's nose. In any case, good personal hygiene usually keeps things from getting out of hand, a seemingly simple solution perpetually undermined by war, famine, and poverty.

Hundreds of mite species are potential agricultural pests capable of causing widespread damage to food crops and ornamentals. The houseplant craze of the 1970s familiarized city dwellers with the family Tetranychidae—spider mites that spin webs as they parasitize plants. They have evolved with specialized mouthparts for tapping the kegs of plant cells. During long summer droughts, entire fields of alfalfa may turn yellow as plants unbathed by rain become heavily infested. Grains and grasses are home to numerous other types that normally are parasites on insects but will attack people.

The ecological complexity of mites is a natural wonder, whether or not we can bear the thought of it, let alone the sight.

• **musk** •

A somehow marvelously onomatapoeic word for one of the oldest ingredients of PERFUME. It is obtained from an abdominal gland in *Moschus moschiferus,* the antlerless deer native to China and Tibet. The term traces back to the Sanskrit *muska,* which means "scrotum" or "testicle," but literally "little mouse." Ancient Persians, masters of the trade, adopted it as *mushk,* which sounds seductive even at this great distance—a "Sabean odor from the spicy shore / Of Araby the blest," wrote Milton in *Paradise Lost.* In the Koran, paradise is populated by "black-eyed nymphs of the purest musk." Horny Greeks pronounced it *móschos,* which became *muscus* in Latin, *musc* in French, and finally *musk* in English at the end of the fourteenth century.

Adult male musk deer, which were hunted nearly to extinction before the advent of synthetic chemicals, secrete the waxy brown gunk as a sexual PHEROMONE. Its active ingredient, called muscone, is a ring-shaped chain of fifteen carbon molecules that is ravishingly large for such compounds. Workable substitutes can be stolen from a few other creatures, such as beavers, alligators, octopuses, and snails, but none of this information seems likely to appear in the advertising bumph on perfume bottles at Saks.

• neutrinos •

Hold out your glove hand like you're going to catch a baseball. Ready? Each second 10 trillion neutrinos will pass right through it. Don't feel klutzy—they'll zip through the Earth, too. And 40 million miles of solid steel.

A veritable gale of neutrinos spewed out by nuclear reactions in the Sun, as well as by the collapse of distant stars and other cataclysmic cosmic events, continually roars through our planet. It has been said that neutrinos are about as close as something can come to being nothing, yet they are one of the most pervasive forms of matter in the universe. To even begin to understand what they are requires letting go of common-sensical notions about the material world.

Neutrino means "little neutral object" in Italian, a most un-Italian concept introduced by the physicist Enrico Fermi (1901–1954) at the University of Rome in 1931. He was a luminary among the international circle of scientists who were hammering out the basic structure of atoms through arcane mathematics and laboratory experiments (a still unfinished endeavor). The existence of neutrinos was deemed a theoretical necessity long before they were confirmed in nature in 1956. So weird did the idea seem at first that Fermi's initial article describing neutrinos was rejected by the British journal *Nature* as being too detached from reality—quite a stern rebuke for anyone in this rarefied line of work.

It is not hard to understand why. Neutrinos are said to have no size, no radius. They might have no mass (an open question, since exactly zero mass is experimentally unprovable). They don't even possess some of the run-of-the-mill abstract qualities that physicists depend on, like electrical charge or electromagnetic force. Yet they exist, more like traveling energy, always on a trajectory at or near the speed of light. The Austrian physicist Wolfgang Pauli (1900–1958) first suggested the existence of such entities (which he called *neutronen* in his native language) in 1930 as the only way to satisfy the principle of energy conservation when an atom ejects an electron from its nucleus in a process called beta decay. (In the jargon of physics, "decay" is when a heavier particle converts into a lighter one.) This was the same year H. L. Mencken opined that "physics itself, as currently practiced, is largely moonshine."

In the late 1920s, scientists had made precise measurements of such radioactive atoms. In beta decay, a neutron was supposed to change into a proton and emit an electron. So they calculated the mass of the initial nucleus, then the mass of the final nucleus plus the energy and mass of the emitted electron using Einstein's $E = mc^2$. The before and after figures should have been equal, but the usual math didn't work properly. Energy was missing somewhere. With great trepidation, Pauli made his audacious suggestion that a small electrically neutral object was carrying the energy away like an invisible bandit. The great Danish physicist Niels Bohr bet him that it would never be found.

Fermi jumped in to help by hypothesizing that the neutron in the nucleus decays into a proton, an electron, and Pauli's little thing. He further posited that a very feeble force in the nucleus is responsible for this reaction. Now universally known as the "weak force," it is, in fact, absolutely wretched compared with electromagnetism—the "strong force" that binds the nucleus together (which, when suddenly released en masse, makes atomic bombs go boom).

Because neutrinos respond exclusively to the weak force, which is felt only at vanishingly short range, the probability that they will interact with any form of matter is vanishingly small. It was estimated that to ensure a collision of one neutrino with surrounding matter would require a target made of lead one light-year thick—beyond even Bill Gates's budget. Fortunately, the development of nuclear reactors in the 1950s provided intense sources of neutrinos. The odds of a collision were much better under these circumstances—though still rare—and resulted in the first experimental evidence of Pauli's prediction, gathered in 1957 by two Americans at Los Alamos, Clyde Cowan and Fred Reines. Neutrinos were not directly observed, of course; rather a chain of events involving subatomic particles was detected that could only be accounted for by their existence. Like true

thieves, their presence is betrayed by something missing—
that is, by the energy they spirit away.

Neutrinos from the Sun were first detected in 1968 using a
100,000-gallon tank of perchloroethylene (dry cleaning fluid)
placed 1,500 meters belowground in South Dakota's Home-
stake Gold Mine. This depth was necessary to screen the tank
from COSMIC RAYS that would cause interactions that mimic
those signaling the presence of a neutrino. When chlorine nuclei
in the fluid captured neutrinos, they gave off electrons and
radioactive argon. Every few months, the tank's contents were
filtered for the argon, a process that extracted perhaps a dozen
argon atoms from the 10^{30} or so atoms in the tank. Although the
apparatus worked like a charm for decades, the number of neutrinos detected consistently amounted to only about half of what
was expected based on knowledge about the Sun, a puzzle veri-
fied by other detectors and still unsolved. Physicists believe that
neutrinos may "oscillate" between three different types, some of
which cannot be detected by current instruments.

Cosmic neutrinos are "seen" by waiting patiently for one to
reveal itself in an extremely large target. To use the whole Earth
as a screen, all detectors are built to look down rather than up.
For example, 10,000 *tons* of ultrapure water were placed in a
Morton–Thiokol salt mine cavern 600 meters beneath the bottom of Lake Erie. The tank was lined with 2,048 light-sensitive
phototubes, which are triggered whenever a neutrino collides
head-on with an electron or proton in the water, creating
charged particles that in turn produce light. These interactions
occur about twice a day, giving a thrill to young wallflower
physicists.

The odds of finding neutrinos were demonstrated clearly in
February 1987, when the closest supernova in 400 years sent
some 60 billion neutrinos through every square centimeter of
Earth facing the ten-second blast. A large detector in Japan cat-

aloged just eleven of them. Japan is now the center of neutrino mania, with its $100 million Super Kamiokande detector holding 50,000 tons of water and 11,242 phototubes in a cavern a mile deep in the Kamioka zinc mine northwest of Tokyo. Because "Super Kam" can also determine the location in the heavens where a neutrino came from, it functions as a type of telescope.

Maurice Goldhaber, a renowned physicist who endured way past the wallflower stage, once said that neutrinos "induce courage in theoreticians and perseverance in experimenters." For the rest of us, they are a good reminder to practice humility.

• noise •

A little noiseless noise among the leaves,
Born of the very sigh that silence heaves.

John Keats mused at one end of the semantic spectrum with his 1816 verse "I Stood Tiptoe Upon a Little Hill," masterfully describing the sound of silence. Today we use this term most often to identify an annoyance, as N. W. McClachlan did in his classic 1935 textbook *Noise:* "sound at the wrong time and in the wrong place."

Noise as unpleasantness traces back to the word's Greek origins, where *nausia* meant "seasickness" or "disgust." In Latin this became *nausea,* whose spelling and meaning are retained in modern English, and then *noise* in Old French, with the sense of commotion and complaint that accompany being seasick. During the nineteenth century, the British navy paid sailors "noise money" to help palliate the maddening blast of foghorns in pea-soup weather, when losing the horizon could also bring heightened propensity for tossing one's cookies. The American wit Ambrose Bierce caught these etymological threads perfectly when he defined *noise* in 1906 as "a stench in the ear."

Unlike the mere handful of noncomparative words available for describing SMELLS, there is a rich vocabulary for noises. The 1963 *Handbook of Noise Measurement* listed 116 common terms: *bang, bark, beep, bellow, blare, blast, blat, bleat, bong, boom, bray, buzz, cackle, cheep, chime, chirp, clack, clang, clank, clap, clatter, click, clink, cluck, clunk, crack, crackle, crash, creak, dingdong, drip, drumming, fizz, glug, gnashing, gobble, grating, grinding, groan, growl, grumble, grunt, gurgle, hiss, hoot, howl, hum, jingle, jangle, kachunk, knock, mew,*

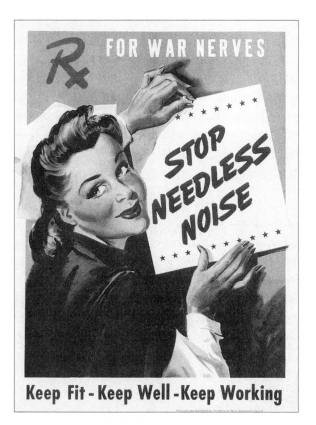

Keep Fit - Keep Well - Keep Working

moan, moo, murmur, neigh, patter, peal, peep, ping, pop, pounding, pow,
pulsing, purr, put-put, rap, rat-a-tat, rattle, ring, rippling, roar, rumble,
rushing, rustle, scream, screech, scrunch, shriek, sizzle, slam, snap, snarl,
snort, splash, sputter, squawk, squeak, squeal, squish, stamp, swish,
swoosh, tap, tattoo, tearing, throb, thud, thump, thunder, tick, tick-tock,
tinkle, toot, trill, twang, twitter, wail, wheeze, whine, whir, whisper,
whistle, yap, yelp, and *zap. Yowl* is inexplicably missing, but this
was before Bob Dylan went electric. *Quel fracas!*

Noise has always been associated with manual labor, at least
until the advent of white-collar work. "Noise is a torture to intellectual people," wrote Arthur Schopenhauer, the nineteenth-
century German philosopher of pessimism, implying that the
lower orders were less distracted by it. He found the cracking
of whips to be an especially irksome sound (a rare annoyance
nowadays outside of fox hunting and S & M circles), despite his
belief in willpower as a fundamental driving force of nature.
Marcel Proust caulked his windows to keep out noise. George
Bernard Shaw ordered his housekeeper never to vacuum, laugh,
or raise her voice when he was at home. Noah Webster, who
sired nine children, built a study with foot-thick walls lined
with cork.

Though the reduction of industrial noise has led to government workplace standards in many nations, some manual occu-
pations have been known to resist the introduction of quiet
machines, such as silent vacuum cleaners (unavailable in Shaw's
day) and typewriters. In a 1927 study, typists performed better
in a noisy office than a quiet one, and said they actually
believed they were working harder in the quiet. This led to the
curious innovation of "acoustic perfume," whereby a little
noise is nice.

Like any other kind of sound, noise is caused by mechanical
vibrations of particles in a gas, liquid, or solid. At room temperature, noise moves through air in waves of pressure at 1,115

feet per second, cork at 1,148 f/s (Webster didn't get much insulation for his money), pure water at 4,723 f/s, hard rubber at 4,875 f/s, sea water at 4,920 f/s, pine wood at 9,460 f/s, concrete at 11,150 f/s, and steel bars or glass at 16,500 f/s. It is well known that noise outside the nominal frequency range of human hearing (20 hertz to 20 kilohertz for young adults, with 1,000 Hz considered the average threshold) can sometimes be bothersome. This is perhaps most obvious in the lower—or infrasound—region, where the rumbling of heavy trucks or the throbbing generated by an open car window at high speeds can actually be sickening. The effect has been exploited by military scientists to design so-called nonlethal infrasound weapons reputed to make enemy soldiers vomit, though combat has been making young men nauseous for thousands of years. On the other hand, low-frequency acoustic vibrations sometimes produce feelings of euphoria.

In fact, human ears are most sensitive to lower pitches—the crucial frequencies for understanding speech lie between 500 and 4,000 Hz. Middle C on a piano is only 261 Hz. The very highest frequency most people can catch is 16 to 18 kHz, whereas chimpanzees can hear up to 22 kHz, dogs up to 38 kHz, cats up to 75 kHz, and bats up to 120 kHz. Nonmammals are generally insensitive to high noises, with birds going deaf at 8 to 12 kHz.

Between the threshold of hearing and the onset of physical pain, the human ear tolerates an enormous range of acoustic energy, differing in magnitude by as much as 10^{14} from low end to high. This energy is measured in units called Bels, first used during telephone cable research at Bell Laboratories to honor founder Alexander Graham Bell: x Bels equals an energy ratio of 10^x between the sound being measured and a standard reference level. Practically speaking, the reference point is the weakest sound that can be heard by someone with excellent hearing in a very quiet place. Thus the human range is 14 Bels, more commonly noted as 140 decibels (1 Bel = 10 dB). Keatsian rustling of leaves is just under 20 dB; an electric shaver buzzes at 50 dB; conversational speech registers between 50 and 60 dB; a toilet flushes at 65 dB; street traffic blares at about 85 dB; shouting hits 90 dB; a jackhammer and live rock music both pummel around 110 dB; and a full concert orchestra playing fortissimo can reach 130 dB. Handheld weapons, such as pistols and rifles, produce noise bursts of 130 dB and up (highest in front of the weapon, where the bullet merits more concern). Standing within twenty-five yards of a jet aircraft engine roaring at 150 dB will definitely hurt, hence those big ear muffs on the ground crew, which attenuate noise by about 50 dB. The Saturn V Moon rocket blasted off at 195 dB, enough to rumble the tummies of spectators on Cocoa Beach miles from the launchpad. At sustained noise levels above 120 dB, industrial workers should be wearing special body protective clothing as well as ear covers. Just stuffing dry cotton in your ears will not provide much protection—a decrease of about 12 dB, at best.

Under everyday circumstances where noise pollution is not a problem, the amount of acoustic power reaching our ears is minuscule. Measured in watts, conversational speech is only 0.00001 W. This produces a displacement of the basilar membranes in our inner ears, which are loaded with sensory cells, on the order of tenths of a nanometer—about the diameter of a hydrogen atom. If damaged, these nerve cells will never grow back.

Technically, loss of hearing sensitivity is divided into three types: *presbycusis,* due to normal aging; *sociocusis,* due to noise exposure; and *nosocusis,* due to disease. By the age of sixty in industrialized societies, half of all men will have suffered a decrease of about 28 dB in their ability to hear frequencies

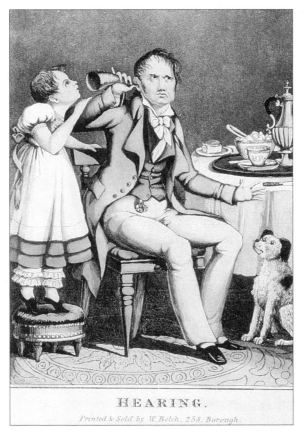

HEARING.

Printed & Sold by W. Belch, 258. Borough.

An evidently serious case of deafness.

around 4,000 Hz, just because of getting older. For women, the loss is about 16 dB. As for sociocusis, it is perhaps unsurprising that the threshold for risk of noise-induced hearing loss is not much above normal speaking volume, at around 75 dB. As such, the deafness of The Who's Peter Townsend and many other vintage rock musicians is easy to comprehend. At the relatively tender age of fifty-one, President Clinton was fitted with hearing aids in each ear due to years of exposure to crowd noise and amplified music. Beethoven, whose sleeve had to be

tugged so that he would notice the cheering after the first performance of his Ninth Symphony, is a famous example of
nosocusis brought on by syphilis.

Recent government studies in Germany found that perhaps
a quarter of the country's youth have suffered significant hearing loss from loud music. Almost 10 percent of student subjects
in Berlin between the ages of sixteen and eighteen had trouble
understanding normal conversation, especially if they used
headphones or went to discos regularly. Officials there want the
European Union to cap the output of personal stereos at 90 dB
(many can put out 120 dB). In 1996, the French legislature
passed a law stipulating a 100 dB max.

In 1627, Francis Bacon recorded his personal experience
with sociocusis: "Standing near one that lured [whistled to call
back a falcon] loud and shrill, [I] had suddenly an offence, as if
something had broken or been dislocated in my ear; and
immediately after a loud ringing (not an ordinary singing or
hissing, but far louder and differing) so as I feared some deafness." In 1831, *The Lancet* published an article by the physician
John Fosbroke titled "Practical Observations on the Pathology
and Treatment of Deafness." The industrial revolution was
well under way, with hazards of continuous occupational exposure to noise common enough for Fosbroke to observe: "The
blacksmiths' deafness is a consequence of their employment; it
creeps on them gradually, in general at about forty or fifty years
of age. At first the patient is sensible of weak impressions of
sound; the deafness increases with a ringing and noise in the
ears. . . . It has been imputed to a paralytic state of the nerve,
occasioned by the noise of forging, by certain modern writers,
and by the old writers, to permanent over-tension of the membrane." Another century would pass before this kind of empirical evidence was substantiated scientifically.

Until recently, the consensus among researchers was that
noise-induced hearing loss occurs when sound vibrations

physically damage the ear. But animal experiments suggest that chemical changes may be the culprit. Noise increases the level of certain enzymes in inner-ear cell membranes, balancing reactions that would otherwise kill sensory cells. This raises the possibility of a deafness "vaccine" for workers exposed to dangerous noise—as well as rock 'n' rollers, of course.

Deafness aside, research has shown that chronic exposure to noise above 90 dB leads to general decline in physical and psychological well-being. In many nations, including the United States, 90 dB is therefore the limit for eight-hour occupational exposure. Just to ensure 90 percent speech intelligibility between two people standing a yard apart, background "white" noise of mixed frequencies cannot exceed 95 dB. *I can't hear you when the water's running!* In Germany, the land of Schopenhauer, workplace guidelines recognize three noise-level maximums: 55 dB for intellectual tasks (learning, intense concentration, creative thinking), 70 dB for uncomplicated office work, and 85 dB for all other kinds of labor. Using an electric shaver in the math library is presumably frowned upon.

Loud digital sound in movies such as *Twister,* the *Star Wars* trilogy, and *Star Trek* (a Starfleet Cruiser can hit 95 dB, even in the vacuum of outer space) triggered a warning from Dolby Laboratories that sound engineers who work regularly on trailers could suffer hearing damage. Theaters typically adjust their speaker systems around 85 dB, the reference point for studio soundtracks, then come down a few to achieve comfortable levels.

Why does a scream make our "blood run cold"? The auditory system has direct connections with the sympathetic nervous system. Sudden intense noises lower our blood pressure and heart rate. The effect is small—not even as great as that accompanying a good laugh—and disappears after a few repetitions (never cry wolf), but it helps alert us to danger. Sound is a much better warning signal than light, so we have evolved with a valuable hair trigger.

Is it any wonder that the ultimate tribute is a moment of silence?

• **no-see-ums** •

This is, as every white person knows, how Native Americans used to talk. *You can't see them?* No see um, *kemo sabe.* The term has been applied to damnably small insect pests since the mid–nineteenth century, but there is still confusion about exactly what it refers to. The usually authoritative *Webster's Tenth* defines it as a "biting midge," which would be fine except that these little flies (family Chironomidae, 1–10 millimeters long) do not bite, though they're often mistaken for mosquitoes when swarming. Perhaps, as the *Random House Unabridged* says, it means punkies (family Ceratopogonidae, 1–5 mm long), even tinier flies whose females sometimes do nip at humans but generally prefer flowers or other insects. Or it could be the dreaded black gnat (*Leptoconops kerteszi,* 1–3 mm long), which is fierce enough to make people vacation elsewhere next year. Maybe what we have here is just a little joke, a bit of levity among Indians who were rapidly running out of things to laugh about as those itchy, scratchy, Manifest Destiny–blinded settlers claimed every last square inch of bug-infested land.

• **ozone** •

A pale blue, poisonous form of oxygen found in trace amounts near the Earth's surface and in a concentrated stratospheric shield. Basically, we need it to go away down here but stay around up there.

The term was coined in 1839 by Christian Friedrich Schönbein (1799–1868), a chemistry professor at the University of Basel, Switzerland, who noticed a peculiar *Gestank* while running electrical currents through water. In German he said *Ozon,* which is derived from the Greek *ózein,* meaning "to smell." His discovery indeed has a distinctive biting odor, sometimes noticeable in a mild way after thunderstorms, when it is generated as ordinary oxygen molecules (O_2) are dissociated by lightning bolts and the free atoms recombine with other O_2 molecules to form triatomic ozone (O_3).

This obnoxious brand of oxygen is highly unstable—it wants to get rid of that extra member and go back to being O_2. It will give away the oxygen atom to almost any other substance it comes into contact with. Ozone is such a powerful "oxidizer" that it can damage many materials, such as rubber, nylon, and the living tissues of plants and animals. Its ability to kill BACTERIA makes it a potent disinfectant, but its character as a severe lung irritant able to trigger asthma, bronchitis, and other respiratory ailments is its most villainous biological role. In Schönbein's day, some scientists were so captivated by ozone that it was promoted as a therapeutic agent for lung disease, much as X RAYS were naively employed for many years after Röntgen's discovery.

Ozone occurs naturally in clean air, but in harmless quantities—just a few tens of parts per billion. Measurements taken near Paris between 1876 and 1910 averaged around 10 ppb. (The most unpolluted regions of Europe today have 20–45 ppb.) Its reputation as a health hazard grew because of so-called photochemical smog, the choking brown haze in which it is produced as sunlight interacts with the hydrocarbons and nitrogen oxides spewed by gasoline engine exhaust. Los Angeles, with abundant sunshine and bumper-to-bumper freeways, was the cradle of ozone pollution. Vegetable crops in the L.A. basin began to show damage in the 1940s. By the 1950s, its bucolic Mediterranean atmosphere had been sullied with acrid ozone concentrations sometimes peaking over 400 ppb. The current federal standard is 120 ppb as a one-hour average, above which occur watery eyes, headache, nausea, reduced lung volume, and bronchial constriction. In other words, you start to go blind and choke to death. Healthy children and adults will experience lung damage even below this level when exercising.

Strict emissions controls have relegated L.A.'s extremes to the sordid history of industrialization, but "ozone action days" to limit tailpipe exhaust and curtail outdoor activity are run-of-the-mill in and around many American cities every summer. Washington, D.C., had seven "Code Red" alerts in 1995, when ozone levels exceeded 125 ppb/hour. (The nation's capital has never satisfied long-term ozone goals since the Clean Air Act was passed in 1970.) A peak of 185 ppb/hour was recorded between Washington and Baltimore during a July 1997 heat wave, the highest since 1988. The mid-Atlantic region is especially bedeviled by weather patterns that import nitrogen oxides and hydrocarbons from coal-burning power plants and other industries as far away as the Ohio River valley. Only about half of Baltimore's ozone is homegrown. With cars the undisputed major culprit, government policies that encourage suburban sprawl and place a higher priority on road improvements than transit lines have exacerbated the problem.

In general, temperatures in the mid-nineties, stationary high pressure, no clouds or WIND, and no rain will set the stage for dangerous buildups. Children, the elderly, and anyone with chronic heart or lung problems should think twice about going outside. Any use of gasoline-powered equipment or vehicles should be postponed; refueling should be done after dusk; use of oil-based paints and solvents or any other household products that release fumes should be deferred. Since ozone takes several hours to form in the sun, peak levels are recorded from 3 to 7 P.M.

Rural areas far from the sources of pollution are no longer safe havens, since the complexities of ozone chemistry permit great rivers of O_3 to form over vast geographic areas. Ozone levels in the Northern Hemisphere are known to be 100 to 200 percent higher than a century ago, rising 1 to 2 percent a year since 1970. Of course, the burgeoning of megalopolises such as São Paulo or Mexico City where environmental regulation takes a backseat to economic development, compounded by biomass burning in the countryside, represents a giant step backward.

There may be no safe haven. On a much smaller—but still irritating—scale, ozone created indoors by copiers, faxes, printers, and other electronic equipment can react with volatile organic compounds from furnishings, paints, dry-cleaned clothes, and even personal deodorants and perfumes to create "office smog." Scores of airborne chemicals mixing in ways that are not yet well understood may harm delicate computer circuitry and cause adverse health effects such as headaches, fatigue, and respiratory problems.

Before the early 1970s, interest in ozone's other important atmospheric domain—a ten-mile-thick layer of 10,000 ppb concentration centered in the stratosphere about twenty-five miles high—was limited to academic circles. British physicist Sydney Chapman (1888–1970) had first proposed its existence in 1930 based on studies of how ULTRAVIOLET rays from the Sun break apart O_2 molecules. The ozone layer is lowest at the poles, highest at the equator, and so rarefied that it would compress at sea level to a thickness of three millimeters. Without its absorption of ultraviolet energy, life could not exist on the planet's surface. (Ozone has also been detected on Ganymede, the Jovian moon, and on two of Saturn's moons, Rhea and Dione. In these instances, O_3 is produced in surface ice by ultraviolet radiation.)

Two developments via the aerospace industry brought this subject to a far wider audience. First, researchers showed that fleets of supersonic transports like the Concorde could inject enough nitrogen oxides into the stratosphere to alter the delicate chemistry of the ozone shield. Second, scientists studying the effects of the space shuttle's solid rocket booster exhaust for NASA found that chlorine atoms could destroy ozone molecules. The shuttle would never be launched often enough to cause problems, but other researchers soon realized that manufactured chlorofluorocarbons (CFCs) such as FREON—used for half a century as cleaning agents, refrigerants, fire suppressants, aerosol propellants, and for other basic industrial needs—were already doing the same dirty deed. In fact, man-made chlorine compounds are the only ones up there in big numbers. Sherwood Rowland, a chemist at the University of California, Irvine, won a Nobel Prize in 1995 for his work connecting CFCs to ozone depletion.

In 1974, the use of CFCs in spray cans was banned in the United States, but worldwide production soon took up the slack. In 1985, the journal *Nature* reported the discovery of an ozone "hole"—an area of reduced ozone concentration—in the stratosphere over Antarctica that could only be accounted for by CFC reactions. (Similar but less drastic depletions have since been found over the Arctic.) Within two years, twenty-four nations signed a treaty requiring them to stop production of most CFCs (as well as three commonly used bromine compounds, which are ten times worse vis-à-vis ozone destruction) by 2000. In 1992, this deadline was advanced to 1996 after a

NASA satellite discovered that chlorine monoxide levels over North America were 50 percent higher than over Antarctica.

Because most chlorine compounds can linger in the atmosphere for a century, ozone levels will keep dropping anyway, allowing more and more UV radiation to reach Earth. According to the U.S. Senate report supporting the Clean Air Act amendments of 1990, this will result in "increased rates of disease in humans, including . . . skin cancer, cataracts, and, potentially, suppression of the immune system." A 1994 United Nations study estimated that every 1 percent decrease in atmospheric ozone causes at least a 1 percent increase in skin cancer incidence.

In 1996, the National Oceanic and Atmospheric Administration found that ozone holes opened earlier in the year and lasted longer than in any period on record. The 12,000-square-mile zones, including a 45 percent decrease in O_3 values over northern Europe, were the worst ever.

Why *do* some people jog on city streets in the noonday sun?

• particulates •

During our stay of three weeks at St. Fago . . . the atmosphere was often hazy, and very fine dust was almost constantly falling, so that the astronomical instruments were roughened and a little injured. The dust collected on the *Beagle* was excessively fine-grained, and of a reddish brown color.

Anyone who's ever been stuck in traffic behind a bus knows as much about particulates as Charles Darwin, who described a Saharan dust fall off the west coast of Africa in 1833. Diesel engines are relatively low emitters of CARBON MONOXIDE, but they belch plenty of black smoke made of tiny carbon particles. These flecks adsorb carcinogenic hydrocarbons onto their large surface areas, creating breathable bullets that lodge in lung tissue. Complaints? Call 1-800-EAT-GRIT.

The size spectrum of particulates is divided by convention along a continuous scale from about 0.001 microns (millionths of a meter; μm) in diameter to 100 μm, the width of a human hair. They can be solid or liquid, natural or man-made, a health threat or just a nuisance. If the atmosphere were totally devoid of them, there would be no rain clouds because the air would be too clean for WATER VAPOR condensation; relative humidity could reach 800 percent. *Dusts* are defined as suspensions of solid particles mixed in a gas by mechanical disintegration, such as crushing and grinding, explosions, or volcanic eruptions. (Airborne dusts are called aeolian, after the Greek god of the WIND. Fine-grain mineral DUST can blow two miles high off the Sahara and eventually give the air over Florida a reddish tinge.) *Smokes* are made of particles from burning or condensation, generally smaller than dusts, that therefore do not settle as quickly. *Fumes* are smokes mixed with noxious vapors. *Mists* are composed of atomized liquid droplets; if concentrated enough to reduce vision, they are *fog;* if made of particles with absorbed water, they are *haze. Aerosols* are clouds of particles in air that are small enough for thermal motion to keep from settling.

At the lower end of size are some metallurgical dusts and fumes whose particles are clusters of molecules smaller than viruses. At the upper end are fine sands, mists, fly ash, coal and cement dusts, SPORES, and POLLEN that may be visible to the naked eye. (Particles larger than 20 μm are heavy enough to settle out of the atmosphere quickly.) In between are such hazards as tobacco smoke (0.01–1 μm), carbon black (0.01–0.15 μm), insecticide dusts (0.15–10 μm), oil smokes (0.015–1 μm), smog (0.01–5 μm), and sundry other waftures.

Particles this small behave like gases. We breathe them, and, depending on their size, they land at various depths in our respiratory system—larger ones in the nasal passages, smaller in the windpipe or lungs. The body's own defense mechanisms can sooner or later clear away inhaled bits larger than about 2.5 μm in diameter, taking anywhere from less than a day to months for a clean sweep. (Even the ruined lungs of coal miners who die of black lung disease may contain less than 10 percent of the dust originally deposited.) Under 2.5 μm is the most dangerous category for human health, though dusts up to 5 μm are still considered damaging to lungs. Tiny particles can carry high concentrations of sulfates, nitrates, and poisonous metals such as mercury, lead, and chromium. The longer they stay in the lungs, the more likely these substances will leach

Donora, Pennsylvania, after the killer smog of 1948.

into other organs, too. Particulates may even amplify the toxic effects of pollutants produced by the same source, such as SUL-FUR DIOXIDE in coal burning, as experienced during the air pollution disasters of Donora, Pennsylvania, in 1948 and London in 1952.

On a global basis, about two-thirds of particulates are natural and one-third man-made. But locally, the natural versus anthropogenic fraction is often reversed. One to 3 trillion (10^{12}) kilograms of dust are normally emitted into the atmosphere every year, yet the 1982 eruption of El Chichon alone released 10^{10} kg of dust (as well as 1.2×10^{10} kg of sulfur compounds and an amount of hydrochloric acid equivalent to 9 percent of the HCl in the whole atmosphere). The "clean" background level of particulates in North America is about 20 micrograms (millionths of a gram; µg) per cubic meter of air. This translates into about 30 million tons of dust particles kicked up across the United States every year by the wind, mostly from soil erosion. In urban areas, the mass concentration may range into the hundreds of micrograms. (A measurement taken in Chicago in 1915 topped 1,000 µg.) Polluted air, moreover, is usually fouled with the smallest particles, which damage health the most.

Coal-fired power plants and engine exhaust account for most man-made particulates. Normally healthy city dwellers may not always be aware of what they're breathing, but they know how grimy their windowsills get even high above street level. American Cancer Society studies have shown that particulates cause more than 64,000 premature deaths every year in U.S. cities from heart and lung disease, with the greatest number occurring in Los Angeles (~5,900), New York (~4,000), and Chicago (~3,500). When their health histories were matched against levels of small particle emissions, residents of the most polluted regions were found to be 17 percent more likely to die than those in the least. Because federal air-quality regulations have focused on particles down to 10 μm, the current standard of 50 μg of particulates per cubic meter is not strict enough to control the critical under-2.5-μm range. (Since the weight of a spherical particle is directly proportional to the cube of its radius, a few coarse ones in a sample will obscure the presence of many fine ones.) Around 10 μm, atmospheric particles are mainly silicon, iron, aluminum, sea salt, and plant matter. At 0.2 μm, they are sulfates (the major component in most urban areas), nitrates, ammonium, organic and lead compounds.

Besides being a respiratory hazard, some particulates can accidently ignite. Anything that will burn in air as a solid bulk may explode as a dust. Flammable dust clouds are usually visible and confined to indoor spaces. For most dusts, there is a certain combination of particle size and concentration that will either self-heat or ignite on a spark or hot surface. Hence the perennial occurrence of grain elevator, sawmill, foundry, and other agricultural or industrial explosions.

• perfume •

The Versailles court of Louis XV, known as "la Cour parfumée," required that a different perfume be worn every day of the week. The king's paramour, Madame de Pompadour, spent 1 million francs to establish a personal perfume bank that would guarantee her perpetual novelty.

If modern researchers are correct in their hypothesis that perfumes were originally conceived not as sexual attractants but to mask one's truly seductive personal odor from unwanted suitors, then La Pompadour's investment at Versailles was, *tant pis,* a grand gesture in the wrong direction.

Nonetheless, dousing oneself with odiferous substances, whether to lure, repel, or even disinfect, is a custom at least as old as the evidence of ointments found in Egyptian tombs. Crusaders brought Arabian distillations back to Europe, where the queen of Hungary commissioned the first modern perfume—a mixture of oils and alcohol—in 1370. Italians dominated the art until Catherine de Medici's private perfumer defected to France, where olfactory commerce grew to industrial proportions. The word *perfume* appeared in English in the 1530s, borrowed from the French *parfum,* which came from Italian *parfumare.* Befitting the relative backwardness of Anglo-Saxons, *perfume* at first referred to the fumes produced when something—incense, one hopes—burned.

Western sophisticates today wear only a hint of fragrance, if any, but there was a time when the crowned heads of Europe sported so much MUSK that their bedroom servants occasionally fainted. Henry VIII preferred his dollop laced with rose. Civet, a secretion from the anal scent glands of a certain Ethiopian cat, and ambergris, a greasy black excretion from sperm whale intestines, were also used in breathtaking concentrations. The curious, yet apt, use of an auditory adjective in the expression "loud perfume"—which dates at least to John Donne's bawdy youth—cannot begin to describe the impact of these precious schmears.

In early modern Europe as in ancient Rome, every garment, bangle, wig, and domestic animal was considered perfumable, at least by the well-to-do. By the late eighteenth century, fashion replaced putrid animal products with sweet flower essences—especially orange blossom, lavender, jasmine, and rose—meant to suggest innocence (a matter of opinion, of course) rather than carnality. The aristocracy was awash in Eau de Cologne (or *Kolnisches Wasser,* depending on geopolitical sentiment), a concoction of rosemary, citrus, and grape first employed to fend off the plague. Napoleon soaked himself with it daily, perhaps to counteract Josephine's stubborn attachment to musk. The association of complex, fantastically expensive perfumes with *haute* couturiers dates to the 1920s, when Chanel and Patou cultivated a luxury market for mass-produced scents.

Though many of the traditional ingredients are still used by perfume manufacturers, albeit in extreme dilution or synthetic form, the modern product is carefully positioned by advertising images that run the gamet from highly sexed (Calvin Klein's Obsession) to bourgeois adventurous (Guerlain's Champs-Elysées) to just plain rich (Givenchy's Organza) to jocky (Polo Sport, the "women's fitness fragrance") to somehow fatherly (Estée Lauder's Pleasures for Men, or is it for labradors?). Since only the most ardent consumer can possibly identify these commodities in the field, they must be intended to enhance the wearer's imagination, not the smeller's. Smelling them may be almost irrelevant, in fact, with the potent linguistic experience of brand name being more important than olfactory messages.

It is well known that the smell of a perfume varies from person to person, depending on interaction with the individual's own B.O. Hence the tankcar loads of supposedly rare fragrances given to shoppers as free samples. Perfumes are volatile chemicals, of course, which means that as they evaporate off the body their odor changes. Those that are somewhat worth their retail price are formulated to achieve a three-tiered effect from the "top note," the initial fresh blast, through the "heart," a more mellow waft, to the "after-fragrance," which may last until next bath. Stale perfume of any pedigree can be as obnoxious as yesterday's socks.

• **pheromones** •

The term *pheromone* was coined in 1959—from the Greek *pherein* ("to transfer") and *hormon* ("to excite")—by German chemists Peter Karlson and Martin Lüscher to describe chemical messages passed between members of the same species, often for courtship. (*Kairomones* attract other species, *allomones* repel them.) Widespread in nature, they are thought to be of relatively minor importance for humans, though the PERFUME industry continues an age-old quest for Love Mist No. 9. "The smell of thy garments is like the smell of Lebanon," King Solomon rhapsodized long before the scientific-industrial age.

There is definitely something in the air, because it is known that groups of women living in close contact—dormitory or barracks mates, family household members, et cetera—may synchronize their menstrual cycles through a still unidentified pheromone in underarm sweat. Nearby male pits, in turn, can influence the period's regularity, bringing long or short cycles into the normal range of 29.5 ± 3 days. Whether substances in human sweat, saliva, and urine known as *androstenes* and *copulines* act as sex pheromones has been debated inconclusively for years. Gourmands may also wonder whether the high level of androstenes in truffles accounts for their popularity.

The existence of MUSK scents in the animal kingdom must have been empirically obvious to the earliest humans, if only because of the usefulness of such knowledge in hunting. Nineteenth-century entomologists began systematic studies of how female moths and butterflies attract mates from miles away. In the 1930s, the German biochemist Adolph F. J. Butenandt (b. 1903), who had earlier isolated the human sex hormones estrone and progesterone, began looking for the female silkworm moth's sex attractant. Twenty years and a half-million sacrificed moths later, the substance was christened bombykol, after the bug's Latin name, *Bombyx mori*. Each moth carries only about 1.2 micrograms (millionths of a gram; μg) of bombykol, quite sufficient to make millions of males come hither when it wafts across detector hairs on their feathery antennae.

The incredible potency of this pheromone and the sensitivity of the detectors can be appreciated from the minimum concentration of bombykol required for a turn-on: 1×10^{-12} μg/ml, or one trillionth of a millionth of a gram per one thousandth of a liter. Imagine what Chanel would charge for every ounce of an analogous eau de cologne.

Recent research has shown that we humans, like reptiles and many other mammals, may have a specialized "erotic nose," or vomeronasal organ (VNO), consisting of a pair of tiny indentations about a centimeter up from the nostrils on either side of the nasal septum. The VNO is clearly present during embryonic development, between the fifth and eighth week of pregnancy, but then regresses and disappears in most adults. First noticed by anatomists three centuries ago, it is usually ignored by modern textbooks (and thus by many plastic surgeons in the course of a nose job). Instead of being wired to the olfactory cortex like our regular schnozola, the cells behind the pits (cigar shaped, of course) hook straight into the hypothalamus and amygdala—unconscious regions that control hormone levels and emotions. So far, however, there is no anatomical proof of this connection. Whether the VNO actually responds to pheromones or is just a vestige of monkey love is a subject of debate. Happily, amputating the noses of adulterers is not as popular today as in Virgil's time. Having located genes in rats that instruct how to make pheromone receptors, researchers are now looking for similar DNA in people.

Animals also use pheromones to communicate fear and alarm. Pregnant mice will abort their fetuses if exposed to urine from male mice other than the fetuses' father. Karl von Frisch (1886–1982), the Austrian zoologist who described the communicative dancing of honeybees, showed how minnows discharge a chemical from cells in their outer skin when injured. This pheromone causes nearby minnows to dart away. Similar pheromones in other fish species trigger differ-ent reactions, such as schooling more tightly or diving to the bottom.

But it is the social insects that demonstrate the most magnif-icent pheromone repertoire in their complex divisions of labor. Frisch's beloved bees create a chemical symphony of more than thirty different molecules. Some are fanned through the hive by beating wings. As far back as 1609, naturalists observed that a single bee stinger will attract others to attack the same site. Ants, too, are renowned for their chemical signals and, unlike bees, use pheromone trails to map newly discovered food or colonies of slaves. In one type of leafcutter ant, a single milligram of trail pheromone could mark a path three times around the Earth.

When Marx and Engels declared in *The Communist Mani-festo* that "the history of all hitherto existing society is the his-tory of class struggles," they were not writing as entomologists, of course. But even the despised cockroach separates high and low individuals according to the chemical composition of a sex pheromone. Researchers at Transylvania University (where else?) found that by manipulating the quantity of three com-pounds that comprise the pheromone, male roaches signal dominance or submission to competing males and to potential female mates.

These chemical systems are by no means restricted to com-plex life forms. The brown algae or kelp familiar to beachgoers depend on pheromones to bring together male and female reproductive cells called gametes. Sea anemones release an alarm substance when attacked that causes others in the colony to withdraw their tentacles and close up. Earthworms under stress add a special chemical to their slimy mucus that not only warns fellow worms but repels predators such as salamanders and birds. (Some hungry snakes, wouldn't you know, actually

prefer the bummer mucus.) Nematodes, barnacles, and BACTE-
RIA all "communicate" with pheromones. Bark beetles that
carry Dutch elm disease release aggregation pheromones to
muster their forces for a mass attack on single trees.

But the heck with insects, angel. *I have perfumed my bed with
myrrh, aloes, and cinnamon. Come, let us take our fill of love until the
morning* (Proverbs 7).

• photons •

Even the minutest molecule of light,
That in an April sunbeam's fleeting glow
Fulfils its destined, though invisible work,
The universal Spirit guides.

What was Shelley on to here in *Queen Mab,* a century before
Einstein? Just versifying the scientific fashion of his day, which
held to Newton's theory (via Galileo) that light was a stream
of "corpuscles," or tiny bodies of matter. The term *molecule,*
which was then a fancy-sounding French word for mass, provided a bit of alliteration.

Shelley was not quite as up-to-date as this poem of 1813
might imply. In 1804, an English doctor named Thomas
Young had performed an experiment showing that light
behaved not like particles but like waves. When he shined a
beam through a narrow slit, a simple image of the opening
appeared against a backdrop. But if he aimed at two parallel
slits, interference patterns formed on the screen just as when
water waves collide. Actually, light could be shown to act like
waves *and* particles, a paradox that lasted until the first decades
of this century.

For most educated observers, it endures. Only physicists have learned to live with it. They say that light consists of photons—not matter particles in the ordinary sense but discrete bundles of energy. How can a stream of bundles behave like a wave? This is a perfectly good question but an intellectual cul de sac. There is no way to account for the wave-particle duality in common-sensical terms. Archimedes' solar mirrors torched Marcellus' fleet in 212 B.C., regardless. Physicists observe nature, then try to write mathematical equations that both fit what they have "seen" and predict the not-yet-observed. If their math corresponds with everyday experience—as is the case with such classic concepts as mass, acceleration, and momentum—then living is easy. But if not, as has been the case since the advent of Einstein's relativity theories, then predictive power becomes the essence of reality. Wave-particle duality is finessed mathematically by saying, yes, light is made of particles, but we cannot determine that an individual photon will hit a target, only the probability.

To fit what had been observed in his Berlin laboratory, the physicist Max Planck realized in 1900 that a photon's energy must be equal to the light's frequency multiplied by some constant of nature, or $E = hf$. He called h the "elementary quantum of action," but it soon became known as "Planck's constant" in his honor. Thus, blue light always has more energy than red light, regardless of intensity. Planck's constant, which would pop up elsewhere in physics, is minuscule—on the order of 10^{-15}—indicating the atomic scale at which quantum effects become significant. This equation was the starting point of a revolution that dethroned classic (Newtonian) physics, which had been so reassuring to human-scale experience.

Physicists also speak of real photons and virtual photons. The former are able to make a Geiger counter (which really does record the presence of discrete bits of matter) click; the latter can carry the influence of electromagnetic fields (see EMF) across empty space (visible light being just a portion of the electromagnetic spectrum). From high-frequency GAMMA RAYS to low-frequency RADIO WAVES, photons (or "quanta") of decreasing energy and zero mass act as force carriers. Similarly, the attraction or repulsion of two charged particles is carried by an exchange of photons. We owe this dizzying picture to the giants of twentieth-century physics: Planck, Paul Dirac, Werner Heisenberg, Enrico Fermi, Wolfgang Pauli, Richard Feynman, to name but a few who attained celebrity status during the evolution of quantum mechanics.

When light strikes certain metals and knocks off electrons to form an electric current, as in a solar cell, the release is called the photoelectric effect. Einstein won a Nobel Prize for describing it in an article published in 1905, taking his cue from Planck to explain how only light of sufficiently high frequency will work. Even though there are lots of photons in a bright red light, each one lacks enough energy to undo an electron. A dim blue light does just fine. Einstein himself was always uncomfortable with the dual nature of photons and light waves. Oddly enough, Planck disagreed with Einstein's momentous article because it represented such a radical departure from what was already known about the wavelike behavior of light. It would take decades of further research to unify the two concepts.

Human eyes are fantastically sensitive to light. The wavelengths they respond to, between 0.4 and 0.7 microns, just happen to coincide with the peak of the sun's spectral output—about 38 percent of the sun's radiant energy is in this band. But we wouldn't want to be much more perceptive. Only five or six photons are needed to activate a retinal nerve cell. At just ten times this sensitivity, we would see dim colors in tiny bursts instead of continuously.

• pollen •

The male gametophyte of plants, holding sperm necessary for reproduction. *Pollen* originally meant "flour" or "mill DUST" in English, and was not used as a botanical term until the mid–eighteenth century. It derives from a Latin word of the same spelling related to *pulvis*—from which came *pulverize* and *powder*—and *polenta,* a type of barley.

WIND-pollinated plants are the main source of pollen ALLER-GENS, which release irritating proteins when they contact mucosal surfaces in our eyes, noses, and throats. Only about 10 percent of all plant species rely on wind to reproduce, but they comprise 90 percent of all plants. Happily, most flowering plants are pollinated by insects or birds and are thus of little concern to sneezy humans. Unhappily, the windy ones produce such enormous numbers of pollen grains that they are essentially inescapable during their active seasons, as hay fever sufferers know all too well. A hundred million grains of pollen fall annually on every square meter of the planet, sometimes hundreds or even thousands of miles from parent plants.

The allergic potential of various plant species is not well characterized, with a few such as ragweed and some grasses being infamous, but many more being suspected only through local anecdotes. Evolutionary puz-

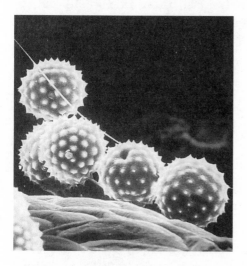

Ragweed, the Genghis Khan of pollens.

zles abound, such as why conifers of the cypress family (which include junipers) are often troublesome, but members of the pine family are usually not, despite the fact that both shed huge tonnages of pollen carried vast distances on the wind. Willow, box elder, and hickory pollen are the strongest tree allergens in North America. Maples and oaks, being so common, no doubt cause more misery.

Grasses, as the most ecologically dominant plants, are prime villains in pollen afflictions, second only to ragweed (which accounts for more cases of hay fever than all other plants together). Bermuda grass, orchard grass, fescue, ryegrass, blue-grass, and timothy—all familiar to country folk and suburban lawnlords alike—are among the most significant allergen pumps. Corn produces extremely allergenic pollen, but because its particles are relatively large and do not waft far, it is bothersome only near the fields.

Perhaps of sentimental interest to aging boomers is the hemp family. Naturalized cannabis is a widespread weed in North America, especially from the Dakotas to Nebraska. Its prodigious pollen output is very buoyant, accounting for as much as 15 percent of the summer ragweed count in Omaha. The uniquely shaped pollen grains produce a buzz all their own: hay fever followed by severe bronchial asthma.

• protozoa •

Single-celled creatures well known to high school biology students as *Paramecium* (first observed in 1674 by Antonie van Leeuwenhoek, who called it the "slipper animacule") and to science fiction fans as *Amoeba* (from the Greek word for change, referring to the shifty green blob that engulfs foolish scientists).

These most complex of microbes cause several diseases in humans, mostly due to poor hygiene or dirty water. For example, amebiasis—from *Entamoeba histolytica*—a colon infection producing mild diarrhea to dysentery, occurs where food and water are contaminated by human feces, such as migrant labor camps and Indian reservations, and among gay male sexual partners. Giardiasis—from *Giardia lamblia*—also travels the fecal-oral route where sanitation is below par. This protozoan has been found in 7 percent of stool samples tested nationwide in the United States. Cryptosporidiosis is a diarrheal disease caused by *Cryptosporidium parvum,* which has triggered mass outbreaks due to water-treatment plant failures, such as the 1993 epidemic in Milwaukee that struck some 300,000 residents. It is life threatening for people with weak immune systems. Toxoplasmosis, spread by *Toxoplasma gondii* in cat feces, infects as much as 80 percent of some populations around the world but is only dangerous for the unborn and the immunocompromised.

Since the 1990 flare-up of civil war in Sudan destroyed that region's medical infrastructure, thousands of people have developed advanced cases of trypanosomiasis, or sleeping sickness, from protozoa of the genus *Trypanosoma* transmitted by tsetse flies. After being limited to less than 1 percent of the population in the late 1980s, incidence is now estimated as high as 40 percent in some populous areas. The disease is capable of inflicting massive fatalities if untreated, as in a 1906 Ugandan epidemic that killed 4 million. Therapeutic drugs like pentamidine cost hundreds of dollars for a complete course, with the price driven up by demand from AIDS patients.

Finally, malaria, one of the great demographic forces in the history of civilization, is caused by four species of *Plasmodium* protozoa. Though essentially gone from the developed world thanks to management of the female *Anopheles* mosquitoes that transmit it, quinine-resistant malaria is considered out of control in tropical Africa, Asia, and Latin America. About 400 million people suffer from it every year, with 1 to 3 million deaths, mostly among children.

Never drink groundwater. If you must venture outside the Hilton bar next time you're in Africa, use that bug spray.

• quarks •

"Three quarks for Muster Mark." James Joyce, the patron fiend of modern English literature, planted this word in *Finnegans Wake* in 1939. He had uprooted it from German, where it referred to a kind of cheese curd and was synonymous with *Quatsch,* which means "trivial nonsense," "absolute rubbish," "bullshit." By the early 1960s, subatomic physics was starting to seem this way even to its high priests, so when Cal Tech theorist Murray Gell-Mann (b. 1929) named a hypothetical new particle the quark, his unscientific term quickly entered the vernacular.

Gell-Mann and his whole profession were burdened with an unwieldy constellation of supposedly "elementary" building blocks of nature, many of which were spawned inside powerful new particle accelerators known popularly as "atom smashers." These huge machines, sometimes miles in length or circumference, were designed to probe the interior of an atom much as one might explore the inside of a peach by shooting it with a pellet at increasing velocities. Eventually the pellet will go fast enough to penetrate the peach's skin and hit the pit. An accelerator imparts so much energy to subatomic collisions, however, that *new* particles that did not exist before the impact are often created from the interaction. (Remember that mass and energy can turn into each other à la $E = mc^2$.)

Since Ernest Rutherford's discovery in 1910 that atoms were not "uncuttable" in the original Greek sense but possessed a structure of nucleus and surrounding electrons, the picture had grown terribly complicated. The use of the word *particle* itself was confusing, since many of the new entities defied ordinary notions about tangible matter. Gell-Mann realized that the hodgepodge of subatomic particles could be organized into groups of eight or ten with similar mathematical properties, rather like chemistry's periodic table. The whimsical (or desperate) physicist called this system "The Eightfold Way" in homage to Buddha's principles for life. He further posited that these groups reflected the existence of an underlying triplet of particles. These became the "three quarks."

From here the story gets even quirkier, or quarkier. Gell-Mann called his three quarks "up, down, and strange." Today three more are recognized: charm, top, and bottom. The six are now referred to as quark "flavors" and considered truly fundamental particles (but still not the only ones), of which protons and neutrons are made, for example. Because they occur only in clusters (e.g., proton = two ups and a down, neutron = two downs and an up), no one has ever seen an individual, or "naked," quark, thus explaining such newspaper headlines as SEARCH FOR BARE BOTTOM.

Quarks are said to be "point-particles," a mathematical concept meaning they have no volume. Moreover, they possess a new kind of charge, similar to an electric charge but occurring in more types than just the traditional positive and negative. It's called "color" but has nothing to do with light. Any intelligent person is entitled to wonder whether quarks are real or just a *façon de parler.* For physicists, this is a false dichotomy. Quarks are obviously nothing like billiard balls, but they're not imaginary. The best answer is that there are collisions observed in atom smashers that can only be explained in terms of quarks.

"What we observe is not nature in itself, but nature exposed to our method of questioning," said Werner Heisenberg (1901–1976), the German physicist whose "uncertainty principle" became emblematic of the field's giddy dictums. (According to this principle, you can determine the velocity of a particle or its position, but not both at once.) Thus, truth is procedural; that is, concepts like quarks are defensible only through their results. This may sound like a logical cop-out to ordinary mortals, but to physicists it is the essence of reality. Anybody who does not experience a kind of vertigo about these subjects, as the Danish physicist Niels Bohr (1885–1962) said, "has not understood the first thing about them."

The dizzying fact is that "things" are not made of many of what physicists call "particles" in the universe. Quark properties have no correlative in the everyday world. Yet they are intimately related to what "things" are made of. Get it? Now try NEUTRINO.

Several times every minute, a quark (a top quark, actually) is really produced somewhere in the Earth's atmosphere by the collision of a COSMIC RAY from outer space with the nucleus of an atom. The quark immediately (truly immediately, in 10^{-24} seconds) decays—that is, transforms—into other particles, leaving behind an unusual arrangement of otherwise normal matter as the only sign of its brief existence. In 1994 researchers at the Fermi National Accelerator Laboratory in Illinois were able to create similar conditions artificially. The top quark turns out to be as heavy as an atom of gold. So quarks are "real," though this word still does not have as much descriptive value to a physicist as to a machinist.

Most physicists now believe that all tangible matter is made of just three particles: electrons, up quarks, and down quarks. If so, these would be the truly "uncuttable" atoms imagined by Greek philosophers 2,500 years ago. The other quarks may have been abundant immediately after the Big Bang. The fact that there is no empirical way to show this does not mean the knowledge is useless. A final answer awaits a final theory of force and matter, which may turn out to say that nature's elementary structures are not "particles" at all.

• **radio waves** •

Radio signals are waves of electromagnetic energy (see EMF) that rise and fall with a frequency from a few thousand cycles per second to a few million million (10^{12}), thus occupying about one-billionth of the total electromagnetic spectrum that extends beyond the 10^{16} million cycles per second of COSMIC RAYS. Practically speaking, the limits of radio frequency for communication are about 10 kHz (thousand cycles per second) to 300 GHz (billion cycles per second). The FBI and CIA control much of the population by transmitting signals that are picked up by tooth fillings or wires implanted in our brains. *Kidding.*

In 1831, the English physicist Michael Faraday (1791–1867), who had left school at the age of fourteen to become an apprentice bookbinder, conducted seminal experiments in electromagnetism during one ten-day spurt of genius. In the early 1860s, the Scotsman James Clerk Maxwell (1831–1879) unified electricity and magnetism with a set of mathematical equations (for all but atomic dimensions, which require quantum mechanics). In 1886, the German physicist Heinrich Hertz (1857–1894) showed that radio waves behave like light waves. Thus the theoretical foundation was laid for technologies that shaped twentieth-century civilization. When Guglielmo Marconi (1874–1937), the Italian son of an Irish woman, became obsessed with wireless communication and demonstrated that radio waves could skip over the horizon by beaming some between St. John's, Newfoundland, and Cornwall, England, in 1901, the global village was born. In 1922 he proposed what would become known as radar, but development of this device depended on war, not commerce.

Any electrical spark, from a household short circuit to a lightning bolt, generates radio waves. To make them carry useful information, such as music, baseball games, or a starlet's come-hither voice, requires shaping the waves with electronic circuitry. Thanks to the insights of Faraday, Maxwell, Hertz, and several generations of clever engineers, this can be done simply and cheaply. Can aliens in outer space hear the Glenn Miller band or news of appeasement at Munich? Well, can they see your flashlight beam when you shine it up into a starry sky? And do the FBI and CIA control the aliens, or vice versa? Much of the television spectrum, at least, is now devoted to answering such important questions.

• radon •

In *De re metallica,* a study of mining published posthumously by the German physician and town administrator Georgius Agricola (1494–1556), silver miners in the Erz (literally "Ore") Mountains between Germany and Czechoslovakia were noted to be so plagued by fatal lung disease that some women there had been widowed seven times. Not until 1879, in an article titled *"Der Lungenkrebs, die Bergkrankheit in den Schneeberger Gruben"* ("Lung Cancer, Mountain Sickness of the Schneeberg

Good old days. In May 1921, Marie Curie receives a gram of radium, the natural source of radon, from President Warren Harding "on behalf of the women of America."

Miners"), was the affliction identified as cancer from autopsies by two German researchers, with mortality rates reaching 50 percent among men in the "death shafts."

As knowledge about radioactivity grew in the first decades of this century, some doctors began to connect these cases with high concentrations of uranium in the ore, but others blamed genetic inbreeding among the insular miners. Digging never slowed down, of course. An Erz mine near Joachimisthal, Bohemia, supplied more than a ton of pitchblende (UO_2) to Marie Curie for her pioneering (and physically backbreaking) isolation of radium. In the United States, radium-containing products—which over the years had included even toothpaste and laxatives—were not banned until 1938. Finally, the burgeoning demand for uranium after World War II provided enough epidemiological evidence to confirm that prolonged exposure to radon gas—a natural radioactive decay product of uranium and radium—causes lung cancer.

Technically, radon (specifically radon-222, the most common isotope) forms directly from ALPHA RAY decay of radium-226, which in turn comes from the uranium-238 that comprises 99.3 percent of all uranium in nature. The word *radon* was coined from *radium emanation* during early studies of radioactivity by Ernest Rutherford, Fritz Dorn, and other physicists around 1900. In their attempts to understand how one substance could emit something that was also radioactive, Rutherford and Frederick Soddy, a chemist, produced the first evidence of elements transmuting into others—the ancient dream of alchemy.

Both uranium-238 and radium-226 are found in most soils and rocks, albeit in widely varying concentrations. Certain types of granite, black shale, and phosphatic rocks are the richest. In the United States, the core rocks of the Appalachians from Maine to Alabama are high in radon, as are areas with granitic sediment in California, Colorado, Idaho, and New Mexico. A radon hot spot surrounds Tampa Bay, Florida, due to phosphate rock high in uranium and radium. But geologic factors alone cannot always predict local radon levels. On average, radon accounts for 67 percent of each American's radiation dose from natural sources.

An invisible, odorless, tasteless inert gas above −143° F, radon leaves the ground and enters surrounding air or dissolves in water. Its concentration depends on how rich the source is and the adequacy of ventilation. Radon itself decays, with a half-life of 3.82 days, into solid polonium isotopes. (Polonium was named by Marie Curie after her native Poland; her maiden name was Manya Sklodowska.) These isotopes, which are also radioactive, used to be called radon daughters but are now correctly addressed as radon progeny. They are actually the main causes of lung cancer.

At the Earth's surface, the health hazard from radon is negligible, but any building will accumulate the seeping gas to some extent—typically four or five times the outdoor concentration. Skin is an effective barrier against alpha radiation, but when radon and its progeny enter the lungs, soft tissue cells lining the airways may be altered by emitted alpha particles to trigger the growth of malignant tumors. (Technically speaking, an alpha particle from polonium-214 will deposit 7.69 million electron volts of energy within about seventy microns of tissue.) Tobacco smoke and radon are known to be synergistic carcinogens. While underground miners were the first to be associated with adverse health effects from breathing radon, the gas is now known to be a ubiquitous indoor air pollutant due to its presence in soil underneath buildings, construction materials like cement blocks, well water, and utility natural gas. In some houses, concentrations have measured as high as those found in mines.

Indoor radon is thus an important public-health problem, though scientists and politicians continue to debate the magnitude of risk. In 1966 the federal government ordered a halt to the practice in some areas of the United States of using uranium mill tailing sands rich in radium-226 to make concrete or as backfill during building construction. In the 1970s more than 700 homes and other structures around Grand Junction, Colorado, were razed or excavated to remove radioactive fill and bricks at an average cost of $13,500. Under the Uranium Mill Tailings Radiation Control Act of 1978, the Environmental Protection Agency established a threshold radiation measure of 4 picocuries (trillionths of a curie, a standard unit of radiation equal to 37 billion disintegrations per second) per liter of air in such houses, above which the government would pay for cleanup.

At this level, the chance of lung cancer from lifetime exposure (seventy years) is equivalent to one to five cases per hundred people, assuming 75 percent of time is spent indoors. In 1984 a house near Boyertown, in Pennsylvania's uranium-rich Reading Prong geologic region, was discovered to have 3,000 picocuries per liter—the equivalent in lung-cancer risk to smoking 135 packs a day—after its owner, Stanley Watras, kept setting off radiation alarms at the Limerick nuclear power plant where he worked. This watershed incident set off a wave of publicity about radon hazards. For comparison, radon levels in Erz Mountain mines averaged about 2,900 picocuries and up.

In 1986, after finding that about 10 percent of the nation's 80 million homes probably harbored radon levels above 4 picocuries, even if they weren't located in uranium-rich regions, the EPA began to issue advisories for the general public. The goal was to make indoor levels as low as those outdoors, which vary geographically—averaging, for example, 5.4 picocuries in New Jersey, 0.8 in Texas, 0.24 in New Mexico, 0.13 in New

York City, 0.12 in Washington, DC. In Maine and Rhode Island, the average radon concentration in private well water exceeds 6,000 picocuries per liter, contributing to indoor pollution through aeration in showers, toilets, and washers. Average indoor exposure in the United States is probably 1 to 2 picocuries.

The EPA estimates that indoor radon exposure causes between 7,000 and 30,000 lung-cancer deaths a year. Anyone wishing to pooh-pooh these figures is free to visit the famous radon spas of Europe, such as Badgastein in the Austrian Alps, where "treatment galleries" boast of 3,000 picocuries. *Prosit!*

• smells •

The "real secret" of class distinctions, wrote George Orwell in *The Road to Wigan Pier,* can be summed up in "four frightful words": *The lower classes smell.* Along these lines, we must also recall the French perfumer of 1709 who proposed a royal scent for the aristocracy, a bourgeois cologne for the middle class, and a disinfectant for the poor. Of course, even Versailles in its heyday reeked of *pisse et merde* not only from court privies and chamber pots, but from every fireplace and dark corner where the ill-mannered relieved themselves regardless of birthright. And pity poor Flaubert, obsessed with the *odeur* of his mistress's slippers.

Being in good odor has always depended on more than fragrance, as *les putains* the world over can attest. If the Devil could mask his sulfurous stench, he would still be evil incarnate. But smells alone have been potent sources of social meaning since ancient times; the Latin antecedent for *sagacious, sagax,* meant having a good sense of smell as well as being perceptive.

The human nose is capable of recognizing tens of thousands of different odors, most with just a single sniff as short as 400 milliseconds. There are upward of half a million different smelly things out there somewhere. In general, women's sense of smell is better by any measure than men's, and they are better at naming them. Yet our main categories are borrowed from a few taste terms: *sweet, burned, pungent, bitter, sour,* and so forth. In the past century of more or less scientific inquiry, researchers have proposed various subdivisions of basic odors, including ambrosial, balsamic, ethereal, resinous, aromatic, putrid, fetid, and goaty. Smells are otherwise designated by what they emanate from. In English there is only one verb to express both inhaling and emitting odors: *to smell.* "You smell; I stink," Samuel Johnson reminded a young lady offended by his sweatiness on a hot London day. While there is an enormous gap between the olfactory environment of past centuries and our deodorized era, the way we talk about smells—in metonymic figures of speech, which refer to the origin of the phenomenon ("like a rose") rather than to a paradigm ("red")—is the same.

Despite this linguistic poverty and the fact that smells have been purged, euphemized, and repressed by modern Western society, their symbolic power has never been eradicated. Hence Proust's search for lost time, his "smells changing with the season, but plenishing and homely, offsetting the sharpness of hoarfrost with the sweetness of warm bread, smells lazy and punctual as a village clock, roving and settled, heedless and provident, linen smells, morning smells, pious smells."

But the difficulty of naming, measuring, preserving, or re-creating smells has ensured that science would give them short shrift compared with other sensory experiences, particularly vision. René Descartes believed unequivocally that the sense of science is sight. Immanuel Kant was downright disparaging: "To which organic sense do we owe the least and which seems to be the most dispensable? The sense of smell. It does not pay us to cultivate it or refine it." Before the twentieth century, what little scientific interest there was in smells derived from their miasmic association with disease. Stench was held either to cause illness, as in malaria (from the Italian *mal aria,* literally "bad air"), or to prevent it, as in the wearing of garlic cloves or folk nostrums against the plague. London cesspools were ordered *open* during Charles II's reign to turn back the plague with their fierce odor. In Swift's *Gulliver's Travels,* the Yahoos treat themselves with a mixture of their own dung and urine, parodying the actual antiplague practice of inhaling stink from a privy hole (preferably on an empty stomach). After Louis Pasteur and Robert Koch established the germ theory of disease in the late nineteenth century, odors depreciated in value, though there are diseases that can be legitimately—albeit archaically—diagnosed from smells (cirrhosis of the liver, gout, yellow fever, typhoid, diphtheria, scurvy, rubella, diabetes).

Only 7 of the 105 natural elements—fluorine, chlorine, bromine, iodine, oxygen (in the form of OZONE), phosphorus, and arsenic—have an odor. Molecules that we can smell must be in a gaseous state and capable of being transported by our inspiration, which normally occurs at a velocity of about five miles per hour. Two branches of the trigeminal nerve—the main conduit for motor and sensory fibers to the face—run from the brain to nerve endings throughout the nasal cavities, where air is screened for irritating odors. Chemical receptors react with a prickly sensation that causes us to sneeze away any offenders. An estimated 10 million olfactory neurons (dogs possess some 200 million) in two narrow chambers behind the nasal bridge are activated when their receptors find matching aromatic molecules in the air we inhale. Smellable substances must therefore also be soluble in the mucus that surrounds these nerve endings. Whether the distinctiveness of their odor is due mainly to chemical characteristics or molecular shape is still not completely understood.

Unlike the neurons for sight, hearing, and touch, our olfactory nerves are connected directly to the brain without any intervening apparatus. The olfactory system has thus long been known as an express train to the brain, as cocaine snorters can attest. The oracle at Delphi, after all, primed herself with wisps of smoldering burel. The fact that the processing of olfactory information is dominated by primitive areas of the brain (amygdala, hippocampus, thalamus) helps explain why the perception of smells is relatively unverbal and has as much to do with emotion as reason. A color will never seem as ugly as a rotten-egg smell is repulsive. PET (positive emission tomography) scans have shown increased activity in the left side of the amygdala, an almond-shaped section of the midbrain associated with fear and other negative feelings, when test subjects smelled sulfurous gases. Regions on the right side of the brain associated with language are not strongly connected to the olfactory circuitry.

Perhaps inspired by Proust, researchers in recent years have tested the tenacity of olfactory memories. As long as laboratory subjects do not have to verbally identify odors, they easily remember many different smells experienced several months earlier. After a year, their scores slip to around 70 percent. This is much better than our retention of visual information, which stays high for the first few weeks and then heads for zero after about four months. But if asked to name the test odors, even our short-term olfactory memory is poor, demonstrating the difficulty most people have when they try to talk about smells. In general, olfactory memory peaks around the age of thirty, like so much else in life.

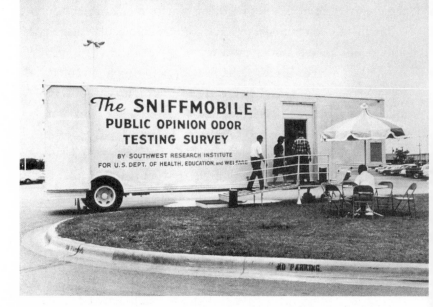

Coming to a city near you. The Sniffmobile, a mobile laboratory for testing public reaction to urban smells.

Such results support the theory that the essential purpose of our sense of smell is to act as an alarm system, provoking quick individual emotional responses rather than elaborate cognitive analyses. *"I smell a rat!"* Almost all odors are immediately liked or disliked; few seem neutral. (For Plato, who was suspicious of the connection between smells and emotions, there were only two classes of odors: stinky and pleasant.) Most people would agree that HYDROGEN SULFIDE is awful, and lavender is pleasant. There are toxic substances, such as CARBON MONOXIDE, that are odorless, as well as some, like cyanide, that are rather nice. And even agreeable odors become disagreeable when they get too strong. Conversely, some stinky ones get delectable at low concentrations. "A dunghill at a distance sometimes smells like musk, and a dead dog like elder flowers," declared

Samuel Taylor Coleridge in an 1823 book titled, oddly, *Table Talk*. In general, preferences and aversions are learned through personal experience, with the latter being more persistent. Only 20 percent of the estimated 400,000 odorous compounds are classified as pleasant, perhaps because people tend to judge any unfamiliar smell as bad.

Artificial olfaction—using arrays of electrochemical sensors connected to computers—has been employed in recent years for quality control in the food and beverage industry, and to monitor the production of smelly chemicals such as airplane deicer. In breweries, for example, electronic noses can detect the presence of diacetyl, which gives beer a buttery "off" flavor. More finely tuned models can discriminate among varieties of wine. Using plastic sensors that swell to different degrees in the presence of various odor molecules, artificial olfaction is mov-

Test sniffers with period hairdos, mid-1960s.

ing closer to the ultimate goal of creating a "nose brain" that consists of thousands of sensors on a single silicon chip along with all the neural network circuitry to run them.

Surveys have found that the most offensive odors from industrial processes are those of hydrogen sulfide, pyridine, butyric acid, phenol, methyl and ethyl mercaptan, dimethyl sulfide, and dimethyl disulfide. In the home, among the worst are carboxylic acids in sweat and rancid food, thiols, amines, and the complex makeup of "stale" tobacco smoke. Tetrahydrothiophene, the natural-gas leak alarm additive that is considered exquisitely smellable, can be detected when concentrations reach about four billion molecules per liter of air. Skunks, by the way, disable enemies up to thirty feet away with a fine spray of butyl mercaptan allomone (see PHEROMONE) shot from two strongly muscled anal glands.

Over the long haul, of course, none of this matters much. Half of all people between the ages of sixty-five and eighty have lost their sense of smell. Past eighty, only a quarter are still getting a good whiff of the world. Of some consolation to men, at least, is the fact that the nose itself doesn't shrink with age and thereby negate one of the last physical emblems of virility.

• **soot** •

A term that traces back to the grimy campfires of prehistory. Its Indo-European base is *sod-* or *sed-*, which are also the roots for "sit" and "settle." Soot is that which settles, namely the fine black particles of nagging concern to any primitive society in constant contact with dirty air.

Like the inhabitants of New York City. Soots are of concern to modern industrial societies because they form during the

THE FOLLY OF IT.

DIRTY OLD MANHATTAN (*to N. Y. Assemblyman*). "Ha, ha, ha! ho, ho, ho! The idea of their asking YOU to help them to clean ME up, when *you* need it a plaguey sight more yourself!"

*Which is sootier, Old Manhattan or the politicians? (*Harper's Weekly, *1881).*

burning of hydrocarbon fuels. Ideally, only CARBON DIOXIDE and water would result from such combustion, but in the real world of engines and furnaces there is also soot (as well as other pollutants, such as CARBON MONOXIDE). Soot forms as fuel molecules agglomerate into carbonaceous particles large enough to be pulled down by gravity instead of riding on the WIND. Besides making a greasy film on city windowsills, they are often carcinogenic. Yes, Virginia, Santa Claus would never survive exposure to so much chimney soot. The chemical industry, which creates lots of it, recycles some as filler in tires, copy machine toner, and printing inks.

Soot occupies a special niche in the study of physics, where it provides a simple means to approximate the ideal of a "black body"—one that absorbs all radiation that strikes it, thus appearing black. Surfaces coated with soot, or carbon black, soak up about 97 percent of incident energy.

———

• **spores** •

Many plants, PROTOZOA, and BACTERIA form special cells that germinate under the right conditions to create the next generation. Think of them as time capsules containing all the information required to restart the creatures' existence. The term *spore* was first used in 1836, borrowed from Greek *spora,* meaning "seed," and *sporos,* "sowing." (As a terminological rule of thumb, spores are asexual, nondividing reproductive cells; POLLEN contains plant sperm; and seeds are embryos for new plants.)

Spores are often remarkable for their tough resistance to extreme environmental conditions and, thus, long-term viability. For reasons that are not completely understood, some can survive incredibly harsh insults: extensive boiling and freezing, radiation exposure, chemical attack, desiccation, and seemingly endless storage. Slime mold spores have been shown to keep on ticking for more than fifty years, while those of mosses may last for thousands. *Bacillus anthracis* bacterial spores can persist for years in alluvial soil like that of the Nile valley, explaining how the Lord could visit "a very severe plague"—anthrax—upon Pharaoh's cattle in Exodus 9. Spores of *Clostridium botulinum* must be heated above 250° F (only possible in a pressure cooker) to guarantee destruction. Such spores, along with their nasty cousins from *C. tetani,* are ubiquitous in nature and present a constant challenge in the prevention of botulism in preserved food and tetanus poisoning in wounds.

Like pollen, plant spores can be allergenic. Bracken fern and club moss are infamous offenders, their spores showing up thickly on air sample slides in the northwestern United States. Dried club moss spores are sometimes used in cosmetic powders, which can cause wicked hay fever, and in fireworks because of their high flammability. (Don't fret about your face catching fire.) Spores from the horsetail rush are in the breeze over the eastern United States from March to September. Each carries four "elaters"—long tendrils with spatulate ends—that expand under dry conditions for extra buoyancy on the wind and hug the spore when wet to help keep it on the ground.

Tropical winds carry as many as 2,800 fungal spores per cubic meter of air across North America in the Arctic. Inhaling those of the fungus *Coccidioides immitis* can cause a disease known as San Joaquin or valley fever. The infection rate may be as high as 90 percent in the U.S. Southwest, though many people never notice it or experience flulike symptoms at worst. Anyone whose immune system is subpar, however, could be in store for a progressive form of the illness with a high fatality rate.

Mushrooms reproduce via spores. In fact, what we call a mushroom is just the aboveground portion of these fungal organisms, a "fruiting body" (somewhat analogous to an apple on an apple tree) that grows expressly for the release of enormous numbers of spores. A four-inch cap can shed 100 million per hour when ripe. If they fall on friendly, unoccupied territory, a new underground vegetative network, or "mycelium," will develop, but if they land somewhere already claimed, they will fuse with the resident mycelium. Most mushroom spores are meant to go airborne, but some rely on novel means of transport. Truffles, for example, need to be eaten by mice or squirrels, which disperse the spores in their feces. Fairy caps hold their spores in a bowl-like depression, from where they are splashed afar by raindrops. And stinkhorns coat their spores with sticky slime, containing the same volatile compounds that waft off a corpse, to attract flies.

In the mid-1970s, mycologists isolated a mycelium of *Armillaria ostoyae* threaded across 1,500 acres in Klickitat County, Washington. The thrill of stomping on a big ripe puffball, *Calvatia gigantea,* comes from watching the exodus of trillions of purple-brown spores.

• sulfur dioxide •

A sharply odored, colorless gas that is powerfully irritating because it turns into sulfurous acid on contact with moisture. Emissions of SO_2 from oil- and coal-fired power plants, petroleum refineries, metal smelters, acid manufacturers, and automotive exhaust have contributed heavily to "acid rain," photochemical smog, and *the destruction of books* in libraries around the world.

Thanks to government intervention, the days are over when a large copper smelting factory, such as in Sudbury, Ontario, denuded its downwind environs for miles of all plant life. But industry still battles federal efforts on behalf of less sulfur dioxide pollution. In 1990, for example, corporate lobbyists warned that new controls on power stations would cost $1,500 per ton of SO_2 removed, driving up the cost of electricity at a rate that consumers would find unpalatable. The actual cost turned out to be $70 to $100.

The U.S. air-quality standard for sulfur dioxide is a maximum concentration of 0.14 parts per million over twenty-four hours. People start to smell and taste it in the range of 0.3 to 1 ppm. In urban areas, hourly peaks may hit 0.5 ppm. Near metal smelters, 1.5 to 2.3 ppm have been recorded. During the catastrophic London smog of December 1952, SO_2 concentrations in air were measured as high as 1.34 ppm. These levels were nowhere near the toxic threshold of 10 ppm, at which healthy people are generally affected (exposure to 0.1 ppm has been shown to bother mild asthmatics), but sufficient in combination with coal smoke to kill from 3,500 to 4,000 Londoners. About 90 percent of inhaled sulfur dioxide is absorbed in the upper respiratory tract, where it plays havoc with mucous membranes, explaining why bronchitis used to be called the national disease of Great Britain.

The fifty largest electric utility companies in the eastern half of the United States account for half of the nation's SO_2 emissions, led in 1995 by Illinois Power. All of the utilities are in compliance with the Clean Air Act, but the law permits regional differences that do not fully recognize how pollution travels. Ever notice any lichens growing near where you live? These hardy symbiotic communities of FUNGI and alga are highly sensitive to SO_2. So if they're gone, beware.

• **thoughts** •

"What can be the traces of words, of actual objects, what further could be the enormous space adequate to the representation of such a mass of material?" wondered Cicero, his head spinning with Roman intrigues, in the *Tusculan Disputations.* Our mind is full of thoughts that no one else can see, all constrained within the convoluted surface of our brain. Since Descartes had the audacity to divide the world into two sepa-

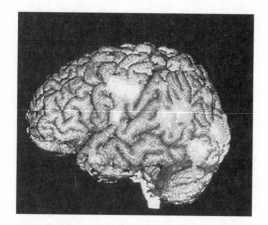

*The waviest gravy: PET and MRI scans
of schizophrenia.*

rate realms, mental and physical, this dualism has probably generated more neurosis than all the mothers in Brooklyn.

Translating the mind's eye into specific neuronal activity is
the Holy Grail of neuroscience. For fifty years, electroen-
cephalography (EEG) has been used to record the electrical
activity that accompanies external stimuli to the brain, but the
technique is dulled by having to place electrodes outside on
the scalp. In the 1970s, X-ray computed tomography, or X-ray
CT, was developed, using precisely focused beams of radiation
that pass through a planar section of tissue at multiple angles to
create an image of the "sliced" surface. This process introduced
a relatively noninvasive way to examine the inside of a living
brain. It was followed by positron emission tomography, or
PET, whereby the brain's blood flow or other chemical functions are portrayed via injected radioisotopes that emit
detectable GAMMA RAYS. Most recently, magnetic resonance
imaging, or MRI, induces atoms to give off RADIO WAVES that
reveal aspects of their biochemical environment. While the latest techniques can show which areas of the brain are working
as it "thinks," they are still too slow and coarsely resolved to

reveal what transpires at the cellular level. Rest assured that when such a device arrives, it will cost more than a penny per thought.

Will happy thoughts look different than other kinds? Peter Pan would never, never take no for an answer. Still, the classic boundary between cognition and anatomy is already gone, at least in the laboratory.

• ultraviolet •

The portion of the electromagnetic spectrum between visible light and X RAYS, first noticed by the British chemist John Ritter in 1801 while studying the effects of sunlight with a prism. It begins just beyond what we can see as violet—hence the Latin prefix *ultra* and the sobriquet "black light." Ultraviolet light is carcinogenic and capable of inflicting severe burns, which is why what used to be called the "California look" is confined to a few living legends like George Hamilton.

Except for occupational or incidental exposure from manmade sources such as arc welders, sunlamps, and mercury vapor or tungsten halide lights, all UV radiation of biological concern comes from the nearest star. Of the 3.9×10^{26} watts of energy given off by the Sun—in terms of wavelengths, from 10 microns (μm) down to 0.2 μm—about 340 watts per square meter reach the top of the planet's atmosphere. Only about 7 percent of this energy is in the form of ultraviolet rays (defined as running from 0.4 μm all the way down to about 0.001 μm), the biggest shares being carried by visible light (from 0.7 μm down to 0.4 μm) and INFRARED.

The most dangerous UVs—those in the so-called vacuum region of the spectrum below 0.2 μm—are absorbed by molecular oxygen (O$_2$) and OZONE (O$_3$) in the atmosphere before reaching the surface. Otherwise, we and every other form of life would be sizzled. "Far" ultraviolet wavelengths between 0.2 and 0.3 μm are also strongly blocked. What little is left shining upon our tender pates—commonly called UVB from 0.28 to 0.32 μm and UVA from 0.32 to 0.40 μm—is not an acute hazard, except to some kinds of BACTERIA (such as, fortunately, *Mycobacterium tuberculosis,* the agent of TB). Certain plant species, like the wild parsnip, depend on these rays to activate chemicals in their leaves and flowers (called furanocoumarins) that poison foraging insects. Busy pollinators such as butterflies and honeybees, which can see ultraviolet light reflected by flowers, are welcome.

UV intensity on any given day varies greatly according to altitude, atmospheric conditions, and the season. Unlike X rays and GAMMA RAYS, ultraviolet radiation is not strong enough to ionize the molecules it hits, so it is less destructive of living tissue. It can't even pass through window glass, so you'll never get tanned in a greenhouse. City dwellers may take small comfort in the fact that smoke and smog also filter it out. But those UVB sunburn rays will buzz right through clouds or fog, as many Cape Cod vacationers discover the hard way every summer. They'll also penetrate a foot into clear water, making water-resistant sunblock a necessity for long days at the beach. Cumulative exposure, amounting to mere hours for a sunburn or many years for cataracts and skin cancer, guarantees regret. Lotions are very effective at keeping everyone happy if rated with SPF > 15 and applied at least a half hour before exposure. If you bake by mistake, oral corticosteroids like prednisone will bring some relief (topical ointments are no better than cold water), but forget anesthetic gunk such as benzocaine.

Ultraviolet irradiation of skin cells has a worrisome ability to distort the DNA helix. It induces the production of chemicals from thymine—one of the four bases encoding genetic information in the DNA chain—that act to impair faithful DNA replication. That is, it may cause mutations. The longer the exposure, the greater the odds of creating cancerous cells—especially in fair, white-skinned persons who started getting lots of sun as children. Due to changes in clothing fashion and leisure time, 1 out of every 84 Americans is now expected to develop melanoma, the least common but most dangerous form of skin cancer, compared with 1 in 600 in 1960 and 1 in

1,500 in 1935. Melanoma is the fifth most common cancer in the United States, with the fastest-growing number of new cases. The risk of developing superficial skin cancers, such as squamous and basal cell carcinoma, is also greatly increased by sun exposure.

Short of malignancy, damaged cells can die and produce the appearance of premature aging. None of the hugely advertised cosmetic "peels" used to turn back the clock of sun damage can reverse such tissue pathology. It is a well-known scientific fact that older women who have invested heavily in peels start to resemble insects, which can see ultraviolet light.

The other major human organ affected by UV is the eye. Reactions range from mild to severe, with conjunctivitis and keratitis (welder's flash or inflammation of the cornea) at one end and cataracts of the lens at the other. Sunglasses that block all UV rays are the preventive answer, particularly at high altitudes or under brightly reflective conditions on snow or open water.

How did that groovy black light in your old dorm room work? Fluorescent lamps produce lots of UV light, which is filtered out by coatings when they are used for conventional lighting. Different coatings will block the visible spectrum but pass UV, making Jimi Hendrix posters glow via induced luminescence of special inks. *Foxy lady!*

• **virus** •

A packet of genetic instructions able to commandeer a living cell's metabolism and make copies of itself. Otherwise, it's as dead as a stone. In Latin the word meant "poison," which expresses with Spartan laconism what these things do to animals, plants, and even BACTERIA. (Here is the perfect place to insert the odd fact that English originally had no words beginning with *v*—they've all been borrowed.) In an age when most bacterial diseases are treatable, viruses are still wild cards of fate.

In the fourteenth century, *virus* referred to any venomous substance. By the early eighteenth, it was being applied to agents of infectious disease, though over a century would pass before anyone understood what was really happening. In 1886, a German chemist, Adolph Mayer, found that whatever was causing destructive mosaic patterns on tobacco leaves could not be cultured. In 1892, a Russian scientist reported that sap from tobacco plants with mosaic disease could infect other plants even after passing through a filter fine enough to remove bacteria. By 1898, tobacco mosaic virus was the first germ to bear the term, followed by the foot-and-mouth disease virus in cattle and the yellow fever virus in humans (1900). With the advent of electron microscopes in Germany during the 1930s, the realm of viruses was opened up to classification.

There are about 5,000 known viruses in a dazzling variety of Buckminster Fullerian shapes from about 0.3 to 0.02 microns long. Several hundred infect humans, mainly via close person-to-person respiratory contact or body excretions. Others prefer animals and infect us only secondarily or if we encroach upon their ecosystems, which may involve insects somewhere in the reproductive cycle. While the human viruses are ubiquitous, the "zoonotic" ones are generally confined to the same environments as their reservoirs and vectors. In most cases of viral infection, there is no specific therapy; your immune system either wins or loses the fight.

Respiratory viruses that cycle primarily in humans include influenza, mumps, Epstein-Barr (mononucleosis), and myriad rhinoviruses of the common cold. Measles is also transmitted in fine droplets spread by coughing or sneezing, but the virus disappears from the nose and throat after a rash appears. Human "enteric" viruses that reside in the intestinal tract and therefore pass around through the so-called fecal-oral route include polio, Coxsackie (meningitis, paralysis), and Norwalk (gastroenteritis). The ability of some viruses to hide from the immune system and lie latent in the body is shown by hepatitis, herpes, and papillomavirus (warts). Retroviruses can cause cancer. Many viruses, like polio, are symptomless more often than debilitating, while a few, like measles and chicken pox, spread like fire in susceptible populations.

Viruses that jump to us from animals or insects include those that cause rabies, encephalitis, yellow fever, dengue, and various exotic hemorrhagic fevers such as Ebola. And, of course, there is HIV, the agent of AIDS. After nearly two decades of frenetic research, no one yet knows where it came from.

Viruses are invulnerable to antibiotics. A handful of antiviral drugs are available, such as amantadine for influenza A and acyclovir for herpes, but most medicines that can zap a virus are

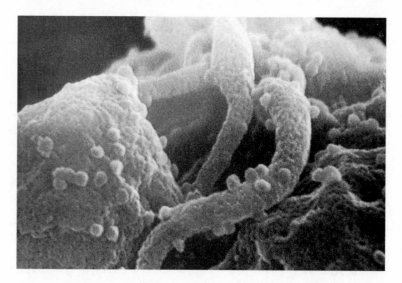

AIDS virus budding from human T-4 immune system cell.

also apt to cause troublesome side effects among good cells. Vaccines are still the only technological adjunct to the natural immune system against viruses, and there are precious few of them in general use. Smallpox remains the single example of a virus—or any other microorganism, for that matter—brought to extinction in nature by the hand of man, though polio has reached that status in the Western Hemisphere.

Since the 1996 "mad cow disease" scare in Britain, subviral particles called prions (pronounced PREE-ons) have entered the medical vernacular, if not many human bodies. An order of magnitude smaller than viruses, prions are believed to be deformed proteins that somehow make normal cell protein molecules mimic their shape. They contain no genetic material and can withstand boiling temperatures, X RAYS, ULTRAVIOLET radiation, and sundry chemical insults, like ten years in FORMALDEHYDE. Whether prions or some more conventional infectious agent cause mad cow disease is subject to debate, though the 1997 Nobel Prize in medicine was awarded to their discoverer, neurologist Stanley B. Prusiner.

A sheep disease known as scrapie (from the intense itching or scraping behavior of infected animals), which has been recognized for centuries, began to appear in British cattle during the 1980s. Characterized by spongy holes in the brain that bring loss of coordination, paralysis, and death, scrapie prions apparently jumped species through feed concentrates containing sheep offal. As vegetarians, cows were totally unprepared to defend themselves against the new pathogens, which also reached zoo antelopes in the same fashion. Over the past decade, they have killed 160,000 cattle in the United Kingdom, with hundreds of thousands more slaughtered as a preventive measure.

A similar form of "transmissible spongiform encephalopathy" (TSE), known as Creutzfeldt-Jakob disease, occurs with extreme rarity in humans throughout the world, mostly adults in their late fifties. In the early 1970s, small studies found that victims were more likely to have dined upon brains. Thus, when sixteen youthful cases arose in Great Britain in 1996, alarms went off about prion-infected meat from mad cows. The feeding of offal to all food animals was banned, suspect herds were destroyed, and exports nose-dived.

American cattle herds are considered to be free of mad cow disease, yet scrapie exists in sheep and mammal remains are not totally banned from feeds. Meatpackers cut away the spinal cord and brain from edible portions of beef, but the common practice of "stunning" cattle with a power-driven plunger through the skull before butchering can spread brain tissue throughout the animal's body. (Poultry, fish, and dairy products have never been implicated in TSE.) Whether and how the disease experienced by animals can vault across the species barrier to humans are matters of intense study. Test-tube experiments have shown that brain protein from cattle with mad cow disease and sheep with scrapie can deform human brain protein. Furthermore, mice injected with brain tissue extracts from human victims of the new variant of Creutzfeldt-Jakob associated with mad cow disease developed symptoms just like those seen in mice exposed directly to extracts from sick cows.

So skipping the *cerveau* might be prudent, along with cutting back on hot dogs and sausages. (There are lots of other good, noncontroversial reasons to do this.) When in rural Kentucky, politely decline offers of fried squirrel brains, which have been implicated in an unusual cluster of Creutzfeldt-Jakob cases there. Or just decide that now's finally the time to go vegetarian.

While it is easy to state that most bacteria are harmless or beneficial, it's a stretch to find something positive about viruses. They do bring new evolutionary possibilities to their host—

human chromosomes may contain remnants of viral DNA. For the sake of diversity, at least, this can be construed as good.

• water vapor •

The Earth's atmosphere contains about 13 quadrillion (10^{15}) kilograms of evaporated H_2O, or water vapor, which is equivalent to a liquid volume of 15,500 cubic kilometers—enough for maybe ten days and nights of global rainfall, not forty. Sounds like quite a bit of wet stuff, but this is only 0.001 percent of the planet's total water supply, about 99 percent of which is in the oceans and polar ice caps.

All that vapor is not evenly distributed, of course. The best-known expression for the amount in local air at a given temperature is *relative humidity,* or the ratio (figured as a percentage) of the existing water vapor pressure—that is, the part of atmospheric pressure caused by water vapor—to the saturation vapor pressure, which is the point where vapor turns into liquid or ice. (*Dew point* is the temperature at which saturation occurs for a given pressure.) Meteorologists know how to measure such parameters, but anyone who spends summer in a big city between New York and Miami learns enough about humidity to avoid repeating the experience if possible.

Finding the weight of water vapor in an imaginary column of air gives another way of comprehending all those quadrillions. By convention, the column covers a one-square-meter patch of ground and rises to the boundary between the troposphere (from Greek *tropos,* "turn," referring to the turbulence of this lowest section of the atmosphere) and stratosphere (from Latin *stratos,* "flat layer")—about 10 kilometers up—where the temperature is cold enough to condense most of the vapor. If all the atmosphere's water vapor were spread out evenly, this pillar would hold about 25 kg. In reality, due to temperature differences, it contains only about 5 kg at the North Pole, increasing to 45 kg at the equator. It averages about 9 kg over the United States in the wintertime and 27 kg in the summer. Describing the humidness of an August day as "heavy" is thus literally correct.

When water evaporates off the oceans or land, the molecules don't stay in the atmosphere very long. After an average of eleven days, they fall back down in raindrops or snowflakes. During this time they may travel long distances on the WIND— a mean path of about 1,000 kilometers in temperate climates. This implies that local precipitation rarely comes from local evaporation. Prodigious amounts of water vapor drift over land from the seas. For example, the summer flow into the United States along the Gulf of Mexico may be ten times the Mississippi's debouchment. But much of it never forms any precipitation at all. Less than two-thirds of the moisture blowing from the Pacific Ocean across the United States and Canada falls to the ground.

Over parched regions such as the Southwest, the fraction may drop to one tenth. Unless water vapor can condense into droplets about 1,000 microns in diameter (ten times the width of a human hair), it won't come down. When the air is supersaturated, moisture still needs PARTICULATES to form droplets around. The most common ones are salt from sea spray, DUST from soil, and pollutants such as sulfuric or nitric acid from fossil fuels (hence, acid rain).

The other major function of water vapor in the atmosphere is to act as a cozy blanket. H_2O molecules in the air are too far apart to interfere with the short wavelengths of solar radiation streaming in from space. But the Earth, being much cooler than the Sun, radiates energy back in long INFRARED waves, which get partially reflected down again by the water vapor. This "greenhouse" action—which is also performed by CARBON DIOXIDE, METHANE, OZONE, and other gases—keeps the surface as much as 54° F warmer than would otherwise be the case.

Where did the planet's water vapor come from originally? Geologists and space scientists have debated for years whether earthbound H_2O derived from primordial volcanic eruptions, as first proposed in the 1890s, or some extraterrestrial source such as meteorites. In 1986, Louis A. Frank of the University of Iowa theorized that comets, whose nuclei are often likened to dirty snowballs, carried water here from the outer reaches of the solar system. His novel idea was held in rather low regard until the spring of 1997, when the NASA Polar satellite detected small comets (about forty feet in diameter, not the twenty-five-mile caliber of Hale-Bopp) entering the atmosphere at a rate of five to thirty every minute. Being so small, they do not heat up to the point where their constituent molecules would break apart. They disintegrate at an altitude of 600 to 15,000 miles, where solar energy vaporizes their ice. This process could account for adding an inch of water to the surface every 10,000 to 20,000 years—quite sufficient to provide all the H_2O now on Earth. In the summer of 1997, the German CRISTA-SPAS atmospheric research satellite launched and retrieved by the space shuttle found evidence of much more water vapor at high altitudes than geological theories can explain.

When Aristotle coined the word we now know as *meteorology* around 330 B.C., he believed that the sun raised up "moist vapors" from the earth, which produced "meteors" of precipitation. Comets, on the other hand, were somehow the result of "dry exhalations" that also caused lightning and earthquakes. Given that he had only four elements to work with—earth, water, fire, and air—this was not bad for seat-of-the-pants cosmology. That comets might turn out to be the source of all moisture would surely have appealed to his imagination, if not his dogma.

• **wind** •

"Blow, winds, and crack your cheeks! Rage!" cries King Lear upon the stormy heath, attended only by his devoted fool. But then there is Shelley's wind among the pines in *Prometheus Unbound,* making "low, sweet, faint sounds, like the farewell of ghosts." Agamemnon sacrificed his daughter, Iphigenia, just to make sure the winds stayed friendly on the way to Troy. Always invisible—or "viewless," as Shakespeare preferred—except for what it perturbs, wind may be the most anthropomorphized of all natural phenomena. From the Aajej, a Morrocan twister, to the Zonda, a hot westerly off the Andes, there are myriad breezes hard and soft, named and sometimes worshiped as though they were sentient beings.

Wind is where the answer is blowing, yet to converse with it is insane. Great doomed empires, like the antebellum South, have utterly disappeared with it. Sowing it creates something far worse. Being three sheets to it is rife with pleasure, but fraught with regret. When it leaves our sails we are done for. And may the gods help anyone who spits or pisses into it.

Some of the earliest graphic symbols, such as the sigma and the swastika, were used by aboriginal residents of North and South America to represent hurricanes. The word itself represents a linguistic whirl over great time and distance from an Indo-European base, *we-,* which became the Greek *aetes,* wind, and *aer,* air, as well as Sanskrit *vatas,* wind, and Russian *vejat,* blow. The progenitor itself eventually turned into *went-,* which evolved into Latin *ventus* (from which English developed *vent, ventilate,* and so forth), prehistoric Germanic *windaz,* and then German and Dutch *wind* plus Swedish and Danish *vind.* The words *breath, spirit,* and *life* are all etymologically entwined through the Greek root *pneuma,* the Arabic *ruh,* and the Latin *anima,* each of which also meant wind. Besides our noun in English, there was once a verb meaning "blow a horn" that survives in the term *wind instrument.* The twists and turns of thought leading to "break wind," the release of FLATUS, should be intuitively obvious.

From the doldrums of equatorial latitudes, named by sailors for their maddening dullness, to the jet streams discovered roaring around the world at 200 miles per hour by World War II bomber pilots, wind or the lack thereof is fundamental to maintaining a liveable planet. "Anybody is as there is wind or no wind there," as Gertrude Stein put it in "An American and France." The 5.6×10^{15} tons of air that envelop the earth, roiled by the planet's rotation, regulate an uneven distribution of solar energy that otherwise would bake or freeze all but a narrow zone passing through San Francisco, Madrid, and Athens in the Northern Hemisphere, and Cape Town and Sydney in the Southern.

Edmund Halley, the English astronomer and comet namesake, first proposed solar heating as the cause of tropical winds in 1680. Immanuel Kant wrote a treatise on wind and an essay about whether the westerly in Europe picked up its moisture over the Atlantic. About 2 percent of all radiant energy from the sun becomes kinetic wind energy, which circulates the air endlessly along lines of equal atmospheric pressure. Alexander Calder's mobiles would otherwise be known as static art. For the continental United States, the mean annual wind speed is only eight to twelve miles per hour, but much higher gusts are ubiquitous. Between 1953 and 1989, 3,550 Americans were killed by some 27,000 documented tornadoes. Windstorms claim an average of 30,000 lives around the world every year.

Possession of a few elementary facts about wind and a bit of open space can make anyone the scientific, if not financial, equal of TV weathermen. With your back to the wind, low pressure is always to the left, high to the right—a rule known as Ballot's Law, after the Dutch physicist who worked it out in 1857. Since the wind turns counterclockwise around a low-pressure area in the Northern Hemisphere, a southerly breeze heralds a storm approaching from the west. (Try drawing it on a piece of paper—it's obvious.) If the wind then shifts or "veers" to the west, the bad weather is heading north and your local skies will probably clear up. If it "backs" to the east, you'll soon be in cold, wet air. By noticing where high clouds are coming from, you can get another reading: with your back to the wind again, if clouds are blowing in from the left, the weather will worsen; from the right, it will brighten. Everything reverses in the Southern Hemisphere, where people walk around upside down anyway.

To further insulate yourself from the inanities of weather infotainment, you can turn the TV itself into a storm detector. (Don't try this unless in the company of a licensed psychiatrist.) Lightning produces electromagnetic waves with frequencies centered on about 150 kilohertz, of which those between 54 and 60 kHz are picked up by televisions. Pop some corn, set the contrast at max, turn to the highest numbered channel, then adjust the brightness down to nearly black. Go back to channel 2 and lightning will make flashes on the tube. If the screen goes completely bright or a picture appears, a tornado may be within twenty miles. Or you may already be in Oz.

Most of us will continue to leave weather forecasting to professionals, but how fast the wind is blowing is an everyday concern. The little spinning cups called anemometers date from the mid-fifteenth century and still provide the bulk of traditional wind speed data, enhanced nowadays by Doppler radar systems that can measure up to 112 mph. (The knot, equal to 1.15 mph, is the standard technical unit.) In 1805, an English naval officer named Francis Beaufort created a scale from zero to twelve corresponding to the effects of rising wind speed on a fully rigged frigate. At force zero (less than one knot), there was not enough wind to go anywhere on a glassy sea. At force twelve (above 65 miles per hour) there was too much to hoist any sail safely. Between force six, a strong breeze of 22 to 27 knots that triggers small craft warnings, and force seven, a moderate gale of 28 to 33 knots, many birds and insects give up flying and humans find that walking straight requires concentration.

Hurricanes are now categorized according to the Saffir-Simpson scale from one (74–95 mph) to five (greater than 155 mph). Between 1899 and 1980, 138 hurricanes battered the continental United States, of which 82 were category 1 or 2, 54 were category 3 or 4, and just 2 were dreaded category 5: the infamous Labor Day storm that struck the Florida Keys in 1935 and Hurricane Camille, which raked the Gulf Coast in 1969, both of which carried maximum sustained winds of about 200 mph. Because friction reduces wind speed to nothing at ground level, hugging the earth is what to do next time you're caught between home and the liquor store in a 'cane.

Tornado wind speeds require their own scale, devised by T. T. Fujita. It was based on analysis of damage produced by a 1970 funnel in Lubbock, Texas, that was strong enough to put a permanent twelve-inch twist in a twenty-two-story steel-frame building. Ranging from F0, where 40-to-72-mph winds cause "light" damage, to F5, where "incredible" destruction results from blasts between 261 and 318 mph (never actually observed, except in Spielberg's dreams), the F-scale indicates that half of all tornadoes carry winds well below 100 mph, which cause only minor structural damage. Australians seem to have special affection for tornadoes, calling them "cockeyed bob" and "willy-willy."

Downtown Lubbock survived the forced abandonment of its contorted office building, but truly Texas-sized skyscrapers cannot be left to the whimsy of Boreas. Each of Manhattan's twin World Trade Center towers rests upon 10,000 shock absorber sandwiches made of two slices of neoprene rubber between three steel plates (what more fitting device in the capital of nosh?). A few miles uptown, the distinctive slanted top of the Citicorp Center conceals an 800,000-pound chunk of concrete that slides back and forth across a steel plate to dampen structural vibrations from high winds. Without such mechanisms, these prestigious monuments to free enterprise might suffer the same ignominy as Boston's John Hancock Tower, half of whose 10,000 sleek blue-green windows had to be boarded up with plywood in the early 1970s until the edifice was bolstered against the breeze.

Since wind pressure increases in proportion to the square of wind speed, it is not hard to understand how great storms can level a landscape. The highest wind gust (nontornadic) ever recorded—231 mph atop New Hampshire's Mount Washington on April 12, 1934—packed a wallop of about 137 pounds per square foot, quite enough to send Mary Poppins skyward.

While he was still a young rake, certainly long before he became dean of St. Paul's, John Donne celebrated the wind's mysteries:

Go and catch a falling star,
    Get with child a mandrake root,
Tell me where all past years are,
    Or who cleft the Devil's foot,
Teach me to hear mermaids singing,
Or to keep off envy's stinging,
        And find
        What wind
Serves to advance an honest mind.

• **x rays** •

"Behind a bound book of about one thousand pages I saw the fluorescent screen light up brightly, the printers' ink offering scarcely a noticeable hinderance." Thus did a German professor named Wilhelm Conrad Röntgen (1845–1923) describe his first observation of the miraculous penetrating power of a strange new radiant energy. The article, dated December 1895, bore the kind of stark title that even a century later conveys the awestruck state of its author: "On a New Kind of Rays." What was the big book? The Bible, perhaps? If so, he was smart not to mention it.

After listing assorted things that were rendered transparent by the mysterious "active agent"—a pack of cards, a sheet of tinfoil, thick blocks of wood, plates of metal and glass, strips of rubber, and so on (one imagines Herr Doktor Röntgen racing around the laboratory trying everything he could lay hands on, including, most importantly, Frau Röntgen)—he ventures to name his discovery *X-Strahlen* in German, or X rays. Here merely was Descartes's traditional mathematical symbol for an unknown quantity, but somehow it captured for the popular imagination a quality both futuristic and sinister.

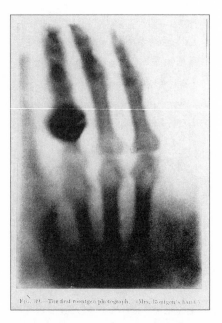

Fɪɢ. 81.—The first roentgen photograph. (Mrs. Röntgen's hand.)

Wilhelm Röntgen's first X-ray photo: Frau Röntgen's hand, mit Ringfinger.

Röntgen had noticed by accident that unexposed photographic plates got foggy after being left near a laboratory fad called a Crookes tube (after British physicist-spiritualist William Crookes, 1832–1919). Similar to modern mercury or sodium street lights, Crookes tubes were glass cylinders through which electrical currents passed in rarefied air, causing eerie fluorescent glows in the residual gas and the glass itself. One night when Röntgen returned to his lab after forgetting to turn off his Crookes tube, he noticed a sheet of paper coated with fluorescent material glowing in the dark near the tube, which itself was covered by cardboard. Surmising that this might be related to the fogged photo plates, he decided that invisible rays must be emanating from the Crookes tube. Lesser researchers had noticed the same effect before, but thought something was wrong with the plates.

The most astonishing feature, by far, was that "if the hand be held between the discharge-tube and the screen, the darker shadow of the bones is seen within the slightly dark shadow-image of the hand itself." Was this a gift from Hades or Mount Olympus?

Röntgen won the first Nobel Prize for physics in 1901. By then, X-ray photographs were a public sensation, though still a scientific conundrum. While researchers tried to figure out exactly where X rays came from, polite society was electrified by the thought that they might reveal a bare bottom under all those Victorian petticoats.

Unlike GAMMA RAYS, which are emitted naturally by radioactive elements, X rays in the Earth's environment must be produced artificially. There are many cosmic sources of X rays, such as the Crab Nebula (the remains of a star that exploded in 1054), but the atmosphere shields us from them. The most common source in everyday life is, of course, the diagnostic or therapeutic X-ray machine, which has been a fixture of medical technology for much of this century. In the years after X rays were discovered, anecdotal evidence—such as dermatitis and hair loss—arose that they could be harmful. Pierre and Marie Curie suffered burns within a year of Röntgen's discovery. Thomas Edison, who foresaw a profitable business in building X-ray machines, quickly abandoned the field after seeing strange pits develop in his assistant's skin and experiencing redness around his own eyes. Initial reports of carcinomas on the hands of radiologists surfaced in 1902. In 1906, after shining X rays on goats' testes, French researchers concluded that reproductive cells were especially vulnerable. But general protective measures were not adopted until 1915 (by the British Röntgen Society). In the 1920s, the ability of X rays to cause genetic mutations was shown in experiments with fruit flies, and birth defects were studied in children whose mothers had been exposed during pregnancy.

Nonetheless, medical X rays were routinely abused as doctors ignored the hazards until the danger of even low-dose exposure was widely appreciated in the 1960s and 1970s. From their earliest application to locate bullets and broken bones, through the use of dyes and radio-opaque barium to photograph internal organs, they are still a basic tool, especially since the advent of CAT (computerized axial tomography) scanners in 1973. Perhaps 90 percent of all nonnatural radiation exposure comes from medical and dental X rays. Benign diseases such as acne and ringworm, which were once zapped copiously, are no longer treated this way. (During the influx of North African immigrants following Israel's independence, doctors there blasted more than 10,000 children with X rays to cure ringworm of the scalp, resulting in a sixfold increase in thyroid cancers.) But sundry forms of cancer are attacked with prodigious doses—aimed (at least in theory) only at tumor cells.

In a typical X-ray machine, a stream of electrons from a heated tungsten filament strikes a small metal dish made of tungsten or chromium, knocking electrons from the inner shells of atoms in the target, which generates X rays. Some of the bombarding electrons are also sharply deflected near atomic nuclei, producing more X rays. The tube in which this occurs is surrounded by heavy lead casing with an aperture through which the useful rays escape. (All X-ray machines leak to some extent, and the beams scatter, which is why your dental hygienist ducks out of the room.) The X-ray beam aimed at teeth or a broken arm is a composite of various wavelengths and degrees of penetrability. Dense tissue like bone absorbs the less energetic rays, leaving shadows on photographic film.

According to the National Academy of Science's BEIR-V report of 1990, an X-ray dose of 300 to 500 rads to the human testes, either acute or over several weeks, can cause permanent sterility. A similar effect on the female ovaries occurs between 250 and 600 rads.

• **zeitgeist** •

Germans love to string nouns together. Here George Wilhem Friedrich Hegel (1770–1831) fused *Geist,* spirit, with *Zeit,* time, to coin a wondrously resonant word for "spirit of the age." Invisible, evanescent, ineluctable, and *stark,* the zeitgeist is not to be trifled with. Just ask the Germans, who followed one of this century's monster zeitgeists straight to hell.

The term was popular at least as far back as the mid-nineteenth century, when Jeremias Gotthelf (1797–1854), a Swiss pastor and moralistic novelist, published a book titled *Zeitgeist und Berner Geist* (roughly, *The Spirit of the Age and the Spirit of Bern*), defending rural values against urban decadence. It eventually acquired the linguistic power of another German word, *Rausch,* which is the feeling one gets from listening to Wagner, say, or the exhilaration of modernism, the druglike "rush" of being perfectly atuned. By now it has largely shed its darksome utopian connotations and become generalized, but it can take them on again in spontaneous fashion.

Perhaps the most curious thing about the zeitgeist is how it coalesces almost before anyone knows it, certainly long in advance of the pundits who describe it. Those enthralled by it are onboard before they know it. There is little to resist should they want to. *Geist* also means ghost, of course, another kind of irresistible ephemeron. And it may be used to express the concept of mind or intellect, the incorporeal part of the brain where THOUGHTS lurk.

What sort of zeitgeist spawns a list of invisible things? These are times of aimless anxiety and greed, poor Reverend God-help might have preached. In the long run, he was on the wrong side of history. Now we can aspire only to a combination of utility and levity for which no one has yet found a word.

· *Bibliography* ·

Agosta, William C. *Chemical Communication*. New York: Scientific American Library, 1992.

Berkow, Robert, ed. *The Merck Manual of Diagnosis and Therapy*. Rahway, N.J.: Merck Research Laboratories, 1992.

Berland, Theodore. *The Fight for Quiet*. Englewood Cliffs, N.J.: Prentice Hall, 1971.

Berner, Elizabeth K., and Robert A. Berner. *Global Environment*. Upper Saddle River, N.J.: Prentice Hall, 1996.

Bockhorn, Henning, ed. *Soot Formation in Combustion*. Berlin: Springer-Verlag, 1994.

Bodanis, David. *The Secret House*. New York: Simon and Schuster, 1986.

Boubel, Richard W., et al. *Fundamentals of Air Pollution*. San Diego: Academic Press, 1994.

Brandt, John C., and Robert D. Chapman. *Rendezvous in Space: The Science of Comets*. New York: Freeman, 1992.

Brookins, Douglas G. *The Indoor Radon Problem*. New York: Columbia University Press, 1990.

Brown, Richard E., and David W. MacDonald. *Social Odours in Mammals*. Oxford: Oxford University Press, 1984.

Calvin, William H. *The Cerebral Code*. Cambridge, Mass.: MIT Press, 1996.

Classen, Constance, et al. *Aroma: The Cultural History of Smell*. London: Routledge, 1994.

Cole, Leonard A. *Element of Risk*. Washington, D.C.: American Association for the Advancement of Science, 1993.

Committee on Science, Engineering, and Public Policy. *Policy Implications of Greenhouse Warming*. Washington, D.C.: National Academy Press, 1992.

Corbin, Alain. *The Foul and the Fragrant: Odor and the French Social Imagination*. Cambridge, Mass.: Harvard University Press, 1986.

Cross, Jean, and Donald Farrer. *Dust Explosions*. New York: Plenum Press, 1982.

Ecker, Martin D., and Norton J. Bramesco. *Radiation*. New York: Vintage, 1981.

Feynman, Richard P. *QED*. Princeton: Princeton University Press, 1988.

―――. *The Character of Physical Law*. Cambridge, Mass.: MIT Press, 1995.

Fishman, Jack, and Robert Kalish. *Global Alert*. New York: Plenum Press, 1990.

Fletcher, Angus. *Colors of the Mind*. Cambridge, Mass.: Harvard University Press, 1991.

Gibbs, Lois Marie, et al. *Dying from Dioxin*. Boston: South End Press, 1995.

Glasstone, Samuel, and Philip J. Dolan, eds. *The Effects of Nuclear Weapons*. Washington, D.C.: U.S. Department of Defense, 1977.

Godish, Thad. *Air Quality*. Chelsea, Mich.: Lewis Publishers, 1988.

Gofman, John W. *Radiation and Human Health*. New York: Pantheon, 1983.

Gregory, Bruce. *Inventing Reality*. New York: John Wiley and Sons, 1990.

Guthrie, George D., and Brooke T. Mossman, eds. *Health Effects of Mineral Dusts.* Washington, D.C.: Mineralogical Society of America, 1993.

Hidy, George M. *Aerosols.* Orlando, Fla.: Academic Press, 1984.

Houck, Marilyn A., ed. *Mites.* New York: Chapman and Hall, 1993.

Huebner, Walter F., ed. *Physics and Chemistry of Comets.* Berlin: Springer-Verlag, 1990.

Kane, Gordon. *The Particle Garden.* New York: Addison-Wesley, 1995.

Kryter, Karl D. *The Handbook of Hearing and the Effects of Noise.* San Diego: Academic Press, 1994.

Lao, Kenneth Q. *Controlling Indoor Radon.* New York: Van Nostrand Reinhold, 1990.

Lederman, Leon. *The God Particle.* Boston: Houghton Mifflin, 1993.

Lederman, Leon, and David N. Schramm. *From Quarks to the Cosmos.* New York: Scientific American Library, 1989.

Lewis, Walter H., et al. *Airborne and Allergenic Pollen of North America.* Baltimore: Johns Hopkins University Press, 1983.

Lippmann, Morton, ed. *Environmental Toxicants.* New York: Van Nostrand Reinhold, 1992.

MacDonald, Gordon J., ed. *The Long-Term Impacts of Increasing Atmospheric Carbon Dioxide Levels.* Cambridge: Ballinger, 1982.

Manchester, William. *A World Lit Only by Fire.* Boston: Little, Brown, 1993.

Martin, Guy, and Paul Laffort. *Odors and Deodorization in the Environment.* New York: VCH Publishers, 1994.

McDaniel, Burruss. *How to Know the Mites and Ticks.* Dubuque, Iowa: Wm. C. Brown, 1979.

Moss, Angela R. *Methane.* Canterbury: Chalcombe Publications, 1993.

National Research Council. *Hydrogen Sulfide.* Baltimore: University Park Press, 1979.

————. *Ozone Depletion, Greenhouse Gases, and Climate Change.* Washington, D.C.: National Academy Press, 1989.

————. *Health Effects of Exposure to Low Levels of Ionizing Radiation, BEIR V.* Washington, D.C.: National Academy Press, 1990.

————. *Rethinking the Ozone Problem in Urban and Regional Air Pollution.* Washington, D.C.: National Academy Press, 1991.

————. *Wind and the Built Environment.* Washington, D.C.: National Academy Press, 1993.

————. *Health Effects of Exposure to Radon.* Washington, D.C.: National Academy Press, 1994.

Peterson, Arnold P. G., and Ervin E. Gross. *Handbook of Noise Measurement.* Concord, Mass.: General Radio, 1972.

Plunkett, E. R. *Handbook of Industrial Toxicology.* New York: Chemical Publishing, 1987.

Pope, Andrew M., et al., eds. *Indoor Allergens.* Washington, D.C.: National Academy Press, 1993.

Posner, Michael I., and Marcus E. Raichle. *Images of Mind.* New York: Scientific American Library, 1994.

Postgate, John. *Microbes and Man.* Cambridge: Cambridge University Press, 1992.

Proctor, Nick H., James P. Hughes, et al., eds. *Chemical Hazards of the Workplace.* New York: Van Nostrand Reinhold, 1991.

Ridley, B. K. *Time, Space, and Things.* Cambridge: Cambridge University Press, 1994.

Rindisbacher, Hans J. *The Smell of Books: A Cultural-Historical Study of Olfactory Perception in Literature.* Ann Arbor: University of Michigan Press, 1992.

Saenz, A. Lara, and R. W. B. Stephens, eds. *Noise Pollution.* Chichester: John Wiley and Sons, 1986.

Schecter, Arnold, ed. *Dioxins and Health.* New York: Plenum Press, 1994.

Sutton, Christine. *Spaceship Neutrino.* Cambridge: Cambridge University Press, 1992.

Tsipis, Kosta. *Arsenal.* New York: Simon and Schuster, 1983.

Turk, Amos, et al. *Human Responses to Environmental Odors.* New York: Academic Press, 1974.

Turner, James E. *Atoms, Radiation, and Radiation Protection.* New York: John Wiley and Sons, 1995.

Vroon, Piet. *Smell.* New York: Farrar, Straus and Giroux, 1997.

Walker, J. Frederick. *Formaldehyde.* New York: Reinhold Publishing, 1953.

Watson, Lyall. *Heaven's Breath.* London: Hodder and Stoughton, 1984.

Wheeler, John Archibald. *A Journey into Gravity and Spacetime.* New York: Scientific American Library, 1990.

Wilkening, M. *Radon in the Environment.* New York: Elsevier, 1990.

Zuckerman, Edward. *The Day after World War III.* New York: Viking Press, 1984.

Sundry news reports from *New Scientist, Nature, Science, Scientific American, Science & Medicine, Outside, The Washington Post,* and *The New York Times.*

· *Index* ·